PRETEST®

Biochemistry

PreTest® Self-Assessment and Review

Eighth Edition

Edited by
Francis J. Chlapowski, Ph.D.
Professor and Chair Ad Interum
Department of Biochemistry and Molecular Biology
University of Massachusetts Medical School
Worcester, Massachusetts

McGraw-Hill
Health Professions Division
PreTest® Series

New York St. Louis San Francisco Auckland
Bogotá Caracas Lisbon London Madrid
Mexico City Milan Montreal New Delhi
San Juan Singapore Sydney Tokyo Toronto

McGraw-Hill

*A Division of The **McGraw-Hill** Companies*

Biochemistry: PreTest® Self-Assessment and Review, Eighth Edition

1 2 3 4 5 6 7 8 9 0 DOCDOC 9 9 8 7 6 5

ISBN 0-07-052089-5

The editors were Gail Gavert and Bruce MacGregor.
The production supervisor was Gyl A. Favours.
This book was set in Times Roman by Compset, Inc.
R.R. Donnelley & Sons was printer and binder.

Library of Congress Cataloging-in-Publication Data

Biochemistry : PreTest self-assessment and review —8th ed./edited by Francis J. Chlapowski.

p. cm.

Includes bibliographical references.

ISBN 0-07-052089-5

1. Biochemistry—Examinations, questions, etc. 2. Clinical biochemistry—Examinations, questions, etc. I. Chlapowski, Francis J.

[DNLM: 1. Biochemistry—examination questions. QU 18.2 B615 1996]

QP518.5.B58 1996

612'.015'076—dc20

DNLM/DLC

for Library of Congress 95-1129

Contents

Preface .... *iv*

Introduction .... *v*

Abbreviations .... *vi*

Amino Acids, Proteins, and Enzymes
Questions *1*
Answers, Explanations, and References *22*

Nucleic Acids
Questions *52*
Answers, Explanations, and References *74*

Carbohydrates and Lipids
Questions *103*
Answers, Explanations, and References *129*

Vitamins and Hormones
Questions *163*
Answers, Explanations, and References *176*

Membranes and Cell Structure
Questions *195*
Answers, Explanations, and References *202*

Metabolism
Questions *208*
Answers, Explanations, and References *218*

Bibliography *232*

Preface

This eighth edition of *Biochemistry: PreTest®Self-Assessment and Review* is based in part on earlier editions prepared by Ian D.K. Halkerston, my colleague at the University of Massachusetts Medical School. Questions were updated to the most current editions of leading textbooks in medical biochemistry. New or expanded questions were substituted when it was felt that recent advances in specific subject areas merited such changes.

Introduction

Each *PreTest® Self-Assessment and Review* allows medical students to comprehensively and conveniently assess and review their knowledge of a particular basic science, in this instance Biochemistry. The 500 questions parallel the format and degree of difficulty of the questions found in the United States Medical Licensing Examination (USMLE) Step 1. Practicing physicians who want to hone their skills before USMLE Step 3 or recertification may find this to be a good beginning in their review process.

Each question is accompanied by an answer, a paragraph explanation, and a specific page reference to an appropriate textbook. In addition, a list of common amino acids begins the chapter "Amino Acids, Proteins, and Enzymes," and a list of regulatory enzymes begins the chapter "Carbohydrates and Lipids." A bibliography listing sources can be found following the last chapter of this text, and a list of abbreviations used in the text follows this introduction.

An effective way to use this PreTest is to allow yourself one minute to answer each question in a given chapter. As you proceed, indicate your answer beside each question. By following this suggestion, you approximate the time limits imposed by the Step.

After you finish going through the questions in the section, spend as much time as you need verifying your answers and carefully reading the explanations provided. Pay special attention to the explanations for the questions you answered incorrectly—but read *every* explanation. The authors of this material have designed the explanations to reinforce and supplement the information tested by the questions. If you feel you need further information about the material covered, consult and study the references indicated.

Abbreviations

ACAT acyl CoA-cholesterol acyl transferase
ACTH adrenocorticotropic hormone
ADP adenosine diphosphate
AMP adenosine monophosphate
ATP adenosine triphosphate
ATPase adenosine triphosphatase
CAP catabolite activator protein
CDP cytidine diphosphate
CMP cytidine monophosphate (cytidylic acid)
CoA coenzyme A
cyclic AMP adenosine 3′,5′-cyclic monophosphate (3′,5′-cyclic adenylic acid)
DHAP dihydroxyacetone phosphate
DNA deoxyribonucleic acid
DNP 2,4-dinitrophenol
DPG diphosphoglycerate
dTMP deoxythymidine monophosphate
dUMP deoxyuridine monophosphate
EF elongation factor
FAD (FADH) flavin adenine dinucleotide (reduced form)
FMN flavin mononucleotide
FSH follicle-stimulating hormone
GDP guanosine diphosphate
GMP guanosine 5′-monophosphate (guanylic acid)
GTP guanosine triphosphate
hCG human chorionic gonadotropin
HDL high-density lipoprotein
HGPRT hypoxanthine-guanine phosphoribosyltransferase
HMG CoA 3-hydroxy-3-methylglutaryl coenzyme A
hnRNA heterogeneous RNA of the nucleus
IDL intermediate-density lipoprotein
IMP inosine 5′-monophosphate (inosinic acid)
IP_3 inositol 1,4,5-triphosphate
LDH lactate dehydrogenase
LDL low-density lipoprotein

LH	luteinizing hormone
mRNA	messenger RNA
MSH	melanocyte-stimulating hormone
NAD (NADH)	nicotinamide adenine dinucleotide (reduced form)
NADP (NADPH)	nicotinamide adenine dinucleotide phosphate (reduced form)
PGH	pituitary growth hormone
P_i	inorganic orthophosphate
PP_i	inorganic pyrophosphate
PRPP	5-phosphoribosylpyrophosphate
RNA	ribonucleic acid
RQ	respiratory quotient
rRNA	ribosomal RNA
TMP	thymidine monophosphate
TPP	thymidine pyrophosphate
tRNA	transfer RNA
TSH	thyroid-stimulating hormone
TTP	thymidine triphosphate
UDP	uridine diphosphate
UMP	uridine monophosphate
UTP	uridine triphosphate
VLDL	very low-density lipoprotein

Notice

Amino Acids, Proteins, and Enzymes

Twenty Common Amino Acids

Name	Type	p*K* of Ionizable Group
ESSENTIAL		
Histidine	basic	6.00
Isoleucine	branched aliphatic	–
Leucine	branched aliphatic	–
Lysine	basic	10.53
Methionine	sulfur	–
Phenylalanine	aromatic	–
Threonine	aliphatic hydroxyl	–
Tryptophan	aromatic	–
Valine	branched aliphatic	–
NONESSENTIAL		
Alanine	aliphatic	–
Arginine	basic	12.48
Asparagine	uncharged polar	–
Aspartate	acidic	3.86
Cysteine	sulfur	8.33
Glutamate	acidic	4.25
Glutamine	uncharged polar	–
Glycine	aliphatic	–
Proline	cyclic aliphatic	–
Serine	aliphatic hydroxyl	–
Tyrosine	aromatic	10.07

DIRECTIONS: Each question below contains five suggested responses. Select the **one best** response to each question.

1. At neutral pH, a mixture of amino acids in solution would be predominantly

(A) dipolar ions
(B) nonpolar molecules
(C) positive and monovalent
(D) hydrophobic
(E) negative and monovalent

2. The figure shown below is

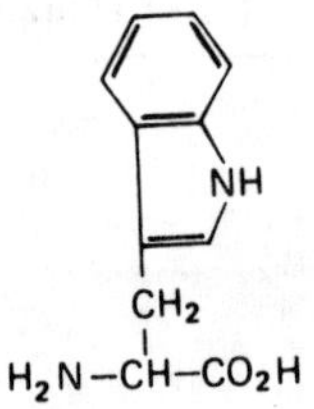

NH

CH_2

$H_2N{-}CH{-}CO_2H$

(A) tryptophan
(B) tyrosine
(C) histidine
(D) phenylalanine
(E) methionine

3. Which one of the following states of hemoglobin (Hb) binds oxygen with the greatest affinity?

(A) Hb
(B) $Hb(O_2)$
(C) $Hb(O_2)_2$
(D) $Hb(O_2)_3$
(E) $Hb(O_2)_4$

4. A comatose man is brought into the emergency room. He is poorly dressed and has a slightly alcoholic smell on his breath. Chemical tests of the patient's blood reveal a blood alcohol level of 5 mg/dL and a blood glucose level of 80 mg/dL. Plasma alanine aminotransferase (ALT, SGPT) levels are normal. In contrast, plasma levels of creatine kinase 2 (CK2), type H lactate dehydrogenase (LDH), and aspartate aminotransferase (AST, SGOT) are considerably higher than normal. What is the diagnosis?

(A) Alcoholic stupor
(B) Myocardial infarction
(C) Hypoglycemia
(D) Cirrhosis
(E) Diabetic hyperglycemia

5. Chronic or long-term regulation of enzymatic activity, as opposed to acute changes in enzymatic activity, is best effected by which of the following regulatory controls?

(A) Covalent modification of enzyme
(B) Synthesis or degradation of enzyme
(C) Product inhibition of enzyme
(D) Availability of substrate to enzyme
(E) Allosteric control of enzyme

6. After a summer of considerable work, a medical student isolates a peptide from a tropical food plant thought to be the cause of the peaceful nature of a group of people inhabiting the forest along a tributary of the Amazon River. Cleavage of the peptide with trypsin yields three fragments. Edman degradation determines the following sequences:

Val-Leu; Ala-Gly-Lys; Val-Met-Arg.

Cleavage of the peptide with cyanogen bromide yields two fragments that have the following sequences:

Arg-Val-Leu; Ala-Gly-Lys-Val-Met.

What is the sequence of the plant peptide?

(A) Val-Met-Arg-Ala-Gly-Lys-Val-Leu
(B) Leu-Val-Arg-Met-Val-Ala-Gly-Lys
(C) Arg-Val-Leu-Ala-Gly-Lys-Val-Met
(D) Val-Leu-Ala-Gly-Lys-Val-Met-Arg
(E) Ala-Gly-Lys-Val-Met-Arg-Val-Leu

---

7. Which of the designations below best describes the relationship of subunits in the quaternary structure of adult hemoglobin?

(A) $(\alpha_1\text{-}\alpha_2)\ (\beta_1\text{-}\beta_2)$
(B) $\alpha_1\text{-}\alpha_2\text{-}\alpha_3\text{-}\alpha_4$
(C) $\alpha\text{-}\beta\text{-}\beta\text{-}\alpha$
(D) $(\beta_1\text{-}\beta_2\text{-}\beta_3\text{-}\alpha_1)$
(E) $(\alpha_1\text{-}\beta_1)\text{-}(\alpha_2\text{-}\beta_2)$

8. Which of the following amino acids is most compatible with an α-helical structure?

(A) Tryptophan
(B) Alanine
(C) Lysine
(D) Proline
(E) Cysteine

9. The isozyme lactate dehydrogenase is composed of two different polypeptide chains arranged in the form of a tetramer. All possible combinations of the different polypeptide chains occur. How many isozyme forms does lactate dehydrogenase possess?

(A) 2
(B) 3
(C) 4
(D) 5
(E) 6

10. A protein chemist is studying solutions of five different polypeptides. The solutions contain a mixture of synthetic peptides, 70 polypeptides long, consisting of repeats of the peptides shown below. Using spectroscopic measurements, the protein chemist notes that in changing the pH of the solutions from pH 5 to pH 8, the configuration of the polypeptide in one of the solutions goes from an α-helical structure to an extended conformation. Which one of the solutions is it?

(A) Leu-Gly-Ala-Met-Gly-Ile-Leu
(B) Asp-Asp-Asp-Asp-Asp-Asp-Asp
(C) Gly-Ile-Lys-Met-Lys-Ile-Lys-Leu
(D) Leu-His-Gly-His-Ala-His-Leu-Leu
(E) Glu-Glu-Glu-Glu-Glu-Glu-Glu

11. Which of the following techniques is used to separate proteins based upon differences in their mass?

(A) Isoelectric focusing
(B) Western blotting
(C) Ion-exchange chromatography
(D) SDS-gel electrophoresis
(E) Dialysis

12. Moonlighting medical students are given the task of trying to find a method for specifically labeling collagen in mice. In one of the sets of experiments they carry out, amino acids labeled with radioactive ^{14}C are added to the diets of mice. They observe that one of the following amino acids is specific in that it does not label collagen. Which one is it?

(A) Serine
(B) Glycine
(C) Aspartate
(D) Glutamate
(E) Hydroxylserine

13. The greatest buffering capacity at physiologic pH would be provided by a protein rich in which of the following amino acids?

(A) Lysine
(B) Histidine
(C) Aspartic acid
(D) Valine
(E) Leucine

14. Transamination of an intermediate of the citric acid cycle results in the formation of

(A) aspartic acid
(B) valine
(C) lysine
(D) serine
(E) alanine

15. Which of the following statements about solutions of amino acids at physiologic pH is true?

(A) All amino acids contain both positive and negative charges
(B) All amino acids contain positively charged side chains
(C) Some amino acids contain only positive charges
(D) All amino acids contain negatively charged side chains
(E) Some amino acids contain only negative charges

16. The concentration of hydrogen ions in a solution is expressed as the pH, which is numerically equivalent to

(A) log $[H^+]$
(B) $-\log [H^+]$
(C) ln $[H^+]$
(D) $-\ln [H^+]$
(E) l/log $[H^+]$

17. The site of enzyme modification by phosphorylation is the amino acid

(A) arginine
(B) cysteine
(C) phenylalanine
(D) lysine
(E) serine

18. Which of the following is a major source of carbon for the one-carbon pool?

(A) Tyrosine
(B) Threonine
(C) Serine
(D) Proline
(E) Glutamic acid

19. The highest concentration of cystine can be found in

(A) melanin
(B) chondroitin sulfate
(C) myosin
(D) keratin
(E) collagen

20. One round of Edman degradation of the peptide H_2N-GLY-ARG-LYS-PHE-ASP-COOH would result in which of the following structures or their phenyl isothiocyanate derivatives?

(A) H_2N-GLY-ARG-COOH + H_2N-LYS-PHE-ASP-COOH
(B) H_2N-GLY-ARG-LYS-PHE-COOH + ASP
(C) H_2N-ARG-LYS-PHE-ASP-COOH + GLY
(D) H_2N-GLY-ARG-LYS-COOH + H_2N-PHE-ASP-COOH
(E) H_2N-ARG-LYS-PHE-COOH + GLY + ASP

21. Enzymes catalyze reactions by

(A) increasing entropy of a system
(B) increasing substrate energy
(C) altering reaction equilibria
(D) lowering total energy levels of reactants
(E) decreasing free energy of activation

22. Analysis of pH 8.6 electrophoretic patterns of hemoglobin isolated from the blood of patients heterozygous for the sickle cell gene would reveal how many bands?

(A) One
(B) Two
(C) Three
(D) Four
(E) Five

23. Since the p*K* values for aspartic acid are 2.0, 3.9, and 10.0, it follows that the isoelectric point (pI) is

(A) 3.0
(B) 3.9
(C) 5.9
(D) 6.0
(E) 7.0

24. When proteins denatured in sodium dodecyl sulfate (SDS) and mercaptoethanol are separated by electrophoresis,

(A) the denatured proteins have a large net negative charge
(B) cations of SDS bind to the protein
(C) disulfide bridges are maintained
(D) the direction of electrophoretic movement is positive to negative
(E) the native protein charge is significant

25. In alkaptonuria (black urine disease), which of the following accumulates abnormally in the urine?

(A) Phenylalanine
(B) Homogentisate
(C) Fumarate
(D) Acetoacetate
(E) Tyrosine

26. In scurvy, posttranslational modification of proteins fails to result in the formation of

(A) glycoproteins
(B) tryptophan
(C) histidine
(D) lipoproteins
(E) hydroxyproline

27. In the cycle shown below, compound A is

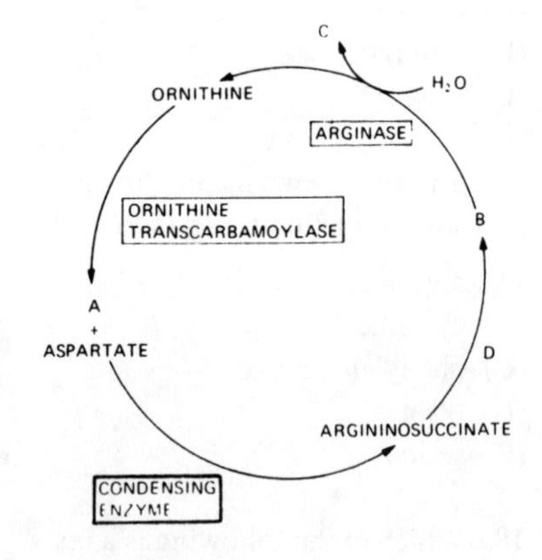

(A) citrulline
(B) urea
(C) carbamoyl phosphate
(D) fumarate
(E) arginine

Questions 28–29

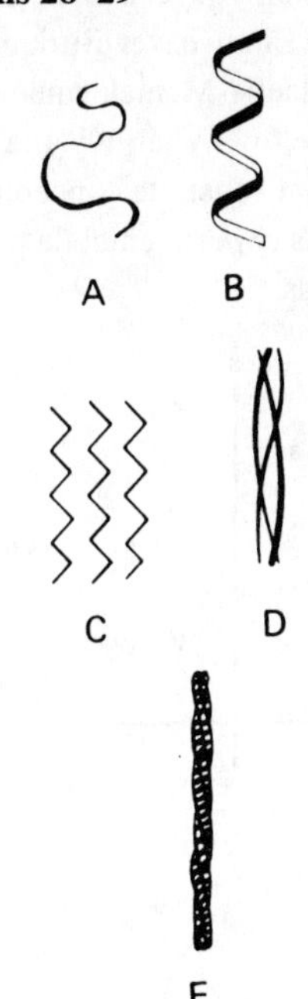

28. Which of the schematic drawings of protein configurations shown opposite represents a super-coiled helix?

(A) Figure A
(B) Figure B
(C) Figure C
(D) Figure D
(E) Figure E

29. Figure D is a three-stranded structure that could represent the conformation of

(A) hemoglobin
(B) α-keratin
(C) polylysine
(D) silk fibroin
(E) tropocollagen

---

30. Contraction of skeletal muscle is initiated by the binding of calcium to

(A) tropomyosin
(B) troponin
(C) myosin
(D) actomyosin
(E) actin

31. According to convention, what is the name of the peptide shown?

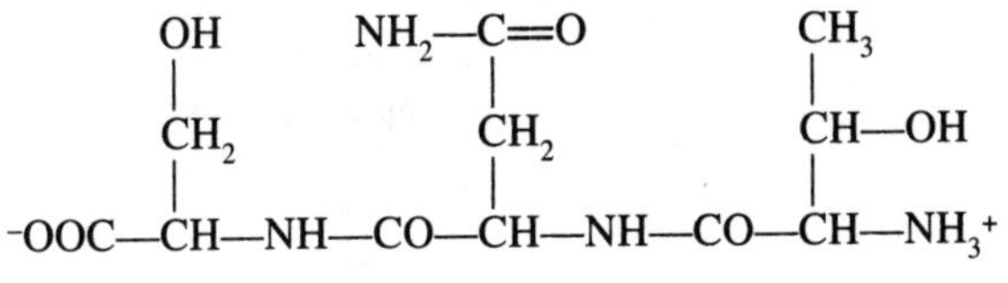

```
      OH        NH2—C=O            CH3
      |              |              |
      CH2           CH2            CH—OH
      |              |              |
-OOC—CH—NH—CO—CH—NH—CO—CH—NH3+
```

(A) Tyr-Val-Ile
(B) Ser-Glx-Thr
(C) Gly-Asn-Thr
(D) Thr-Glx-Ser
(E) None of the above

32. Acid hydrolysis of a peptide reveals equimolar amounts of lysine, glycine, and alanine. Following trypsin digestion of the peptide, only free glycine and a single dipeptide are observed with paper chromatography. In the following examples, the *N* terminal of the peptide is written to the left of the *C*-terminal end. What is the primary structure of the original peptide?

(A) Gly-Lys-Ala-Lys-Gly-Ala
(B) Gly-Lys-Ala
(C) Lys-Gly-Ala
(D) Ala-Lys-Gly
(E) Ala-Gly-Lys

33. Which of the following statements about the peptide bond is true?

(A) It is a carbon-carbon bond
(B) It has *cis* hydrogen and oxygen groups
(C) It is planar
(D) It has rotational freedom
(E) It is found in fatty acids.

34. The normal brown-red color of feces results from the presence of

(A) stercobilin
(B) urobilinogen
(C) bilirubin
(D) mesobilirubin
(E) biliverdin

35. K_m and V_{max} can be determined from the Lineweaver-Burk plot of the Michaelis-Menten equation shown below. When *V* is the reaction velocity at substrate concentration *S*, the *x*-axis experimental data are expressed as

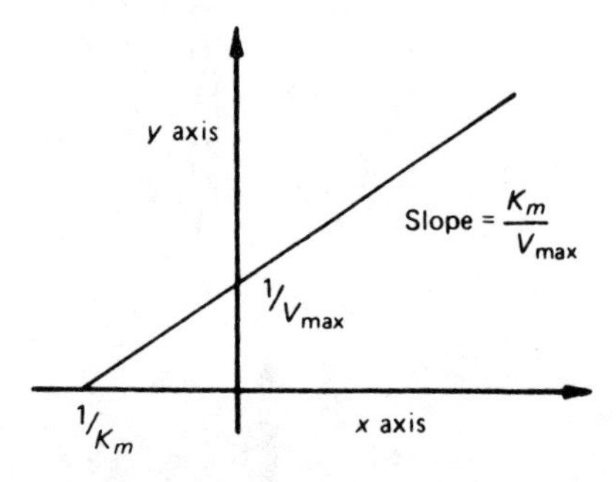

(A) 1/*V*
(B) *V*
(C) 1/*S*
(D) *S*
(E) *V/S*

36. Isozymes for a given reaction

(A) have different substrate specificities
(B) have identical affinities for the same substrate
(C) bind at least two different substrates
(D) exhibit different electrophoretic mobilities
(E) contain similar ratios of different polypeptide chains

37. A purely competitive inhibitor of an enzyme

(A) increases K_m without affecting V_{max}
(B) decreases K_m without affecting V_{max}
(C) increases V_{max} without affecting K_m
(D) decreases V_{max} without affecting K_m
(E) decreases both V_{max} and K_m

38. Which of the following compounds serves as a primary link between the citric acid cycle and the urea cycle?

(A) Malate
(B) Succinate
(C) Isocitrate
(D) Citrate
(E) Fumarate

39. Given that $\Delta G° = -2.3\ RT \log K_{eq}$, determine the free energy of the following reaction assuming the values shown are equilibrium concentrations:

A	+	B	$\rightleftharpoons$	C
10		10		10
moles		moles		moles

(A) $-9.2\ RT$
(B) $-4.6\ RT$
(C) $-2.3\ RT$
(D) $+2.3\ RT$
(E) $+4.6\ RT$

Questions 40–41

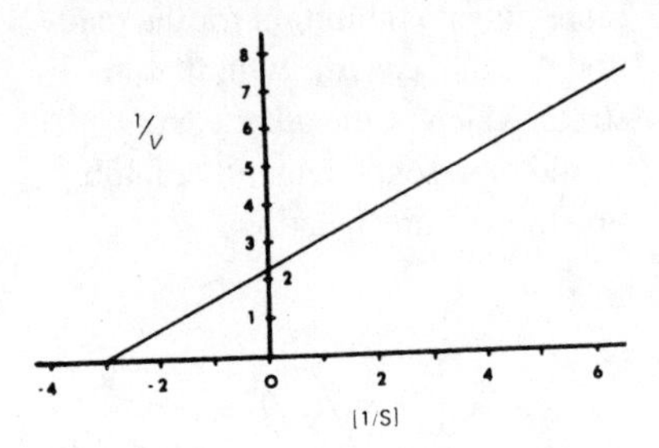

40. The K_m of the enzyme with the kinetic data shown above is

(A) −0.50
(B) −0.25
(C) +0.11
(D) +0.25
(E) +0.33

41. The V_{max} of the enzyme with the kinetic data shown above is

(A) reciprocal of the absolute value of the intercept of the curve with the *x* axis
(B) reciprocal of the absolute value of the intercept of the curve with the *y* axis
(C) absolute value of the intercept of the curve with the *x* axis
(D) slope of the curve
(E) point of inflection of the curve

42. In the study of enzymes, a sigmoidal plot of substrate concentration ([*S*]) versus reaction velocity (*V*) may indicate

(A) Michaelis-Menten kinetics
(B) myoglobin binding to oxygen
(C) cooperative binding
(D) competitive inhibition
(E) noncompetitive inhibition

43. If curve X in the graph below represents no inhibition for the reaction of some enzyme with its substrate, which of the other curves would represent competitive inhibition of the same reaction?

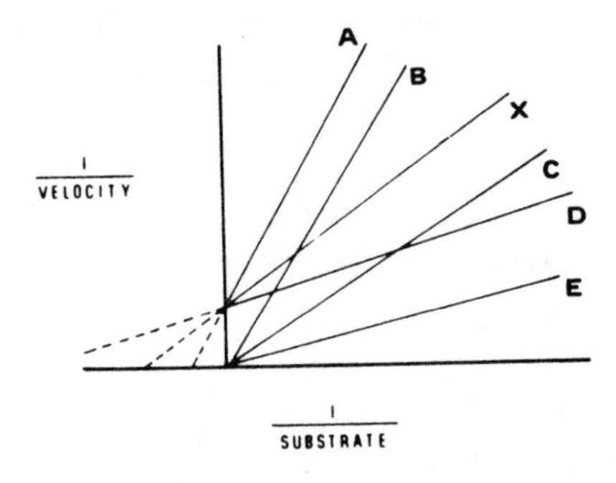

(A) A
(B) B
(C) C
(D) D
(E) E

44. Immunoglobulin G molecules can be characterized by which of the following statements?

(A) They are maintained at a constant level in the serum
(B) They contain nucleic acids
(C) They contain mostly carbohydrate
(D) They can be separated into subunits with reducing agent and urea
(E) They can be separated into subunits with proteolytic enzyme and urea

45. Which of the following techniques for purification of proteins can be made specific for a given protein?

(A) Dialysis
(B) Affinity chromatography
(C) Gel-filtration chromatography
(D) Ion-exchange chromatography
(E) Electrophoresis

46. The reactions of the urea cycle occur

(A) in the cytosol
(B) in the mitochondrial matrix
(C) in the mitochondrial matrix and the cytosol
(D) only in lysosomes
(E) in peroxisomes

47. Depletion of α-ketoglutarate during hyperammonemia leads to the formation of

(A) glycine
(B) arginine
(C) proline
(D) histidine
(E) glutamine

48. A solution of glutamic acid is titrated from pH 1.0 to 7.0 by the addition of 5 mL of a solution of 1 *M* NaOH. What is the approximate number of millimoles of amino acid in the sample ($pK_{a1} = 2.19$, $pK_{a2} = 4.25$, $pK_{a3} = 9.67$)?

(A) 1.5
(B) 3.0
(C) 6.0
(D) 12.0
(E) 18.0

49. The approximate isoelectric pH (pH_I) for aspartic acid ($pK_{a1} = 2.09$, $pK_{a2} = 3.86$, $pK_{a3} = 9.82$) is represented by which of the following values?

(A) 2.09
(B) 2.97
(C) 3.86
(D) 6.84
(E) 9.82

50. The thyroid hormone thyroxine (T_4) is derived from

(A) threonine
(B) tyrosine
(C) thiamine
(D) tryptophan
(E) tyramine

51. Which one of the following statements correctly describes transport of O_2 by hemoglobin?

(A) O_2 binds to hemoglobin more avidly than does CO
(B) The binding of O_2 to hemoglobin causes a valence change in the iron of the heme moiety
(C) Each of the four heme moieties binds O_2 independently
(D) The plot of percentage of O_2 bound versus O_2 pressure is sigmoidal in shape
(E) Increased CO_2 concentrations increase O_2 affinity

52. The velocity-substrate curve below characterizes an allosteric enzyme system. The curve demonstrates that

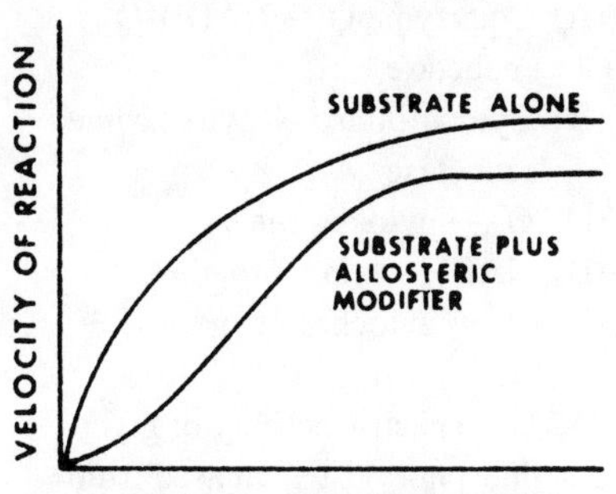

(A) a modifier changes the binding constant for the substrate but not the velocity of the reaction
(B) a modifier binding to the allosteric site can also affect the catalytic site
(C) binding of the substrate is independent of its concentration
(D) binding of the modifier is independent of its concentration
(E) binding of substrate to the allosteric site displaces modifier

53. Which one of the following amino acids can be directly dehydrated and deaminated without the use of aminotransferase enzymes?

(A) Valine
(B) Alanine
(C) Serine
(D) Aspartate
(E) Arginine

54. Which one of the following steps of electron transport is thought to be coupled to phosphorylation of ADP to form ATP?

(A) Coenzyme Q → NADH-Q reductase
(B) Cytochrome *c* → cytochrome reductase
(C) O_2 → cytochrome *c*
(D) NADH → flavoprotein
(E) O_2 → cytochrome oxidase

55. The specific activity of glycogen phosphorylase increased from 2.5 units/mg homogenate protein to 325.5 units/mg protein after being bound to and eluted from a cation exchange column at pH 2.7. What can you conclude from this information?

(A) The yield of enzyme was greater than 80 percent
(B) The enzyme was negatively charged at pH 2.7
(C) The enzyme was purified over 100-fold
(D) The enzyme was globular in structure
(E) The enzyme was in an activated state

56. Which one of the following can be converted to an intermediate of either the citric acid cycle or the urea cycle?

(A) Tyrosine
(B) Lysine
(C) Leucine
(D) Tryptophan
(E) Aspartate

57. Which one of the following proteins is found in the thick filaments of skeletal muscle?

(A) α-Actinin
(B) Myosin
(C) Troponin
(D) Tropomyosin
(E) Actin

58. Which one of the following statements correctly describes allosteric enzymes?

(A) Effectors may enhance or inhibit substrate binding
(B) They are not usually controlled by feedback inhibition
(C) The regulatory site may be the catalytic site
(D) Michaelis-Menten kinetics describe their activity
(E) Positive cooperativity occurs in all allosteric molecules except hemoglobin

59. Which one of the following proteolytic enzymes is activated by acid hydrolysis of the proenzyme form?

(A) Trypsin
(B) Chymotrypsin
(C) Elastase
(D) Pepsin
(E) Carboxypeptidase

60. Which one of the following enzymes is regulated primarily through allosteric interaction?

(A) Chymotrypsin
(B) Pyruvate dehydrogenase
(C) Glycogen phosphorylase
(D) Glycogen synthase
(E) Aspartate transcarbamoylase

61. Which one of the following structures may be classified as a hydrophobic amino acid at pH 7.0?

(A) $^{+}H_3N—CH_2—COO^{-}$
(B) $^{+}H_3N—CH_2—CH_2—CH_2—COO^{-}$
(C) $^{-}OOC—CH_2—CH(^{+}NH_3)—COO$
(D) $^{+}H_3N—CH_2—CH_2—CH_2—CH_2—CH(^{+}NH_3)—COO^{-}$
(E) $CH_3—CH(OH)—CH(^{+}NH_3)—COO^{-}$

62. The figure below shows the structure of immunoglobulin hydrolyzed by papain to form two A fragments and one B fragment. It is true of fragment A that it

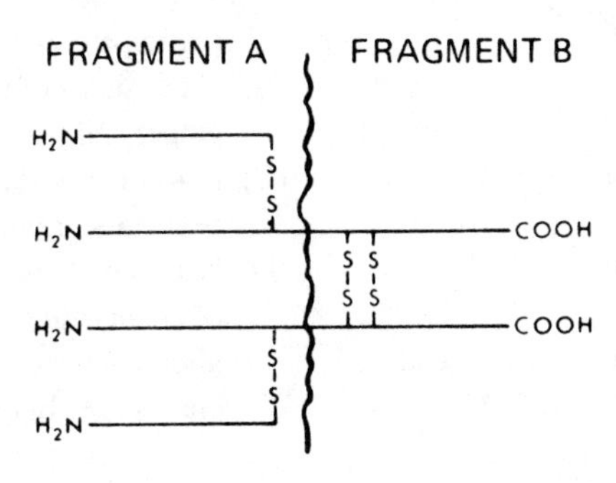

(A) contains the constant regions
(B) is the heavy chain
(C) contains the light chain
(D) is not functional as an antibody-combining site
(E) cannot be further dissociated by mercaptoethanol

63. Which one of the following statements correctly describes allosteric enzymes?

(A) Regulatory molecules bind the active site
(B) Regulatory molecules alter equilibrium but not activity
(C) Regulatory molecules do not affect activity or equilibrium
(D) Hyperbolic plots are obtained when reaction velocity is plotted against substrate concentration
(E) Binding of substrate to one site can affect other sites

DIRECTIONS: Each numbered question or incomplete statement below is NEGATIVELY phrased. Select the **one best** lettered response.

64. All the following statements describe characteristics of the peptide bond EXCEPT

(A) it has a *trans* configuration
(B) it is polar but uncharged
(C) it forms a single bond directly linking the α-carbons of adjacent amino acids
(D) it demonstrates no rotation around the bond
(E) it is planar

65. Tyrosine is a precursor of all the following EXCEPT

(A) dopa
(B) dopamine
(C) norepinephrine
(D) epinephrine
(E) phenylalanine

66. All the following are considered anabolic activities EXCEPT

(A) inhibition of gluconeogenesis in the liver
(B) increased glucose transport into muscle
(C) increased amino acid transport into muscle
(D) increased glycogen formation by the liver
(E) increased lipolysis in adipose tissue

67. All the following amino acids are found in proteins EXCEPT

(A) threonine
(B) homocysteine
(C) cysteine
(D) methionine

68. In comparing the secondary structure of proteins, it can be seen that all the following descriptions apply *only* to a β-sheet and not to an α-helix EXCEPT

(A) all peptide bond components participate in hydrogen bonding
(B) N terminals of chains are together and parallel
(C) it is composed of two or more segments of polypeptide chain
(D) N-terminal and C-terminal ends of chains alternate in an antiparallel manner
(E) chains are almost fully extended

69. All the following amino acids contain only methyl or methylene groups as constituents of their side chains EXCEPT

(A) valine
(B) alanine
(C) glycine
(D) leucine
(E) isoleucine

70. The graph below shows a titration curve of a common biochemical compound. All the following statements about the graph are true EXCEPT that

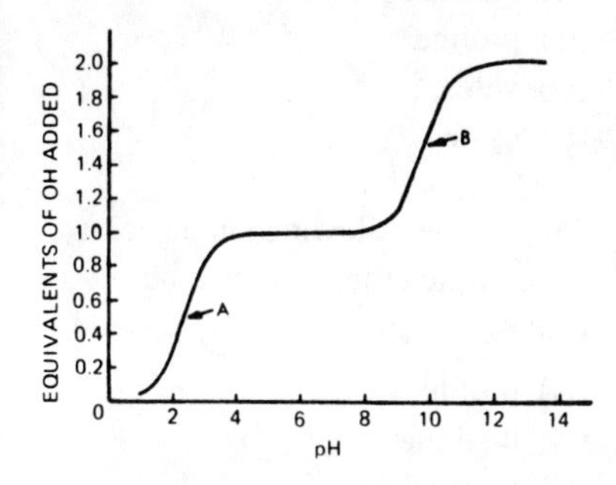

(A) the compound has two ionizable functions
(B) the compound has no ionizable side chains
(C) the maximum buffering capacity of the compound is between pH 5 and 7
(D) point A could represent the range of ionization of a carboxyl function
(E) points A and B represent the respective p*K*'s of an acidic and basic function

71. All the following amino acids are ionizable in proteins EXCEPT

(A) leucine
(B) histidine
(C) terminal glycine
(D) arginine
(E) cysteine

72. All the following are important in determination of the primary structure of a protein EXCEPT

(A) determination of the amino acid sequence in peptide fragments
(B) determination of the extent of α-helix formation
(C) determination of the number of polypeptide chains
(D) cleavage of pure polypeptide chains to smaller chains
(E) separation of different polypeptide chains

73. All the following are considered major metabolic intermediates formed by amino acid degradation EXCEPT

(A) α-ketoglutarate
(B) succinyl CoA
(C) fumarate
(D) oxaloacetate
(E) citrate

74. All the 20 common amino acids are optically active EXCEPT

(A) lysine
(B) aromatic amino acids
(C) glycine
(D) cysteine
(E) serine

75. All the following statements concerning immunoglobulin are true EXCEPT

(A) the distinctiveness of the heavy chains gives the different classes of immunoglobulins their unique biologic characteristics
(B) IgG is the principal antibody in the serum
(C) the light chains are similar in each class of immunoglobulin
(D) the constant regions of the heavy chains are the same in each class of immunoglobulin
(E) IgA is the major immunoglobulin found in external secretions

76. Inactive zymogens (proenzymes) are precursors of all the following gastrointestinal enzymes EXCEPT

(A) ribonuclease
(B) trypsin
(C) chymotrypsin
(D) carboxypeptidase
(E) pepsin

77. In humans all the following amino acids are nonessential EXCEPT

(A) arginine
(B) tyrosine
(C) proline
(D) valine
(E) cysteine

78. All the following amino acids have cyclic groups as their side chains EXCEPT

(A) proline
(B) histidine
(C) tryptophan
(D) tyrosine
(E) aspartate

79. All the following circulatory factors are involved in the clotting process in some manner EXCEPT

(A) platelets
(B) fibrinogen
(C) pepsinogen
(D) plasmin
(E) heparin

80. All the following amino acids are found in unusually high amounts in collagen as compared with other proteins EXCEPT

(A) glycine
(B) isoleucine
(C) hydroxyleucine
(D) proline
(E) hydroxyproline

81. Thalassemias are produced by mutations that cause all the following EXCEPT

(A) deleted α-hemoglobin chains
(B) deleted β-hemoglobin chains
(C) defective β-hemoglobin chains
(D) impaired RNA processing
(E) sickle cell anemia

82. All the following are allosteric enzymes EXCEPT

(A) aspartate transcarbamoylase
(B) phosphofructokinase
(C) hexokinase
(D) pyruvate kinase
(E) lactate dehydrogenase

83. All the following assumptions apply to Michaelis-Menten kinetic analyses of enzyme action EXCEPT

(A) the total enzyme concentration studied at each substrate concentration is fixed in analysis of enzyme kinetics
(B) formation of enzyme-substrate complex does not appreciably decrease the concentration of substrate
(C) K_m decreases with competitive inhibition
(D) maximal velocity is reached when the enzyme-substrate complex is equal to the total concentration of enzyme present
(E) the initial reaction of velocity should be measured since most of the substrate has not been converted to product

84. All the following are glycosaminoglycans EXCEPT

(A) collagen
(B) hyaluronic acid
(C) chondroitin sulfate
(D) heparin
(E) keratan sulfate

85. All the following statements correctly describe properties of proteins EXCEPT

(A) trichloroacetic acid precipitates and denatures proteins
(B) 4 *N* HCl solubilizes most proteins
(C) solubility is highest at the isoelectric point
(D) ammonium sulfate usually precipitates proteins while preserving biologic activity
(E) gel chromatography separates proteins based upon size

86. All the following correctly describe the active site of an enzyme EXCEPT

(A) it is small relative to the entire enzyme
(B) specificity is defined by the arrangement of certain atoms
(C) it is two-dimensional in structure
(D) it is usually a crevice or cleft
(E) it initially binds substrates by weak attractions

87. Under normal conditions in blood, all the following amino acid residues of albumin are charged EXCEPT

(A) arginine
(B) aspartate
(C) glutamine
(D) glutamate
(E) histidine

88. The substitution of valine for glutamate at position 6 on the two β-chains in sickle cell hemoglobin causes all the following EXCEPT

(A) decreased electrophoretic mobility at pH 7.0
(B) decreased solubility of deoxyhemoglobin
(C) polymerization of deoxyhemoglobin
(D) altered primary structure
(E) brittle red blood cells

89. All the following statements describing protein structure are true EXCEPT

(A) many proteins have no quaternary structure
(B) charged amino acid side chains tend to be on the outside of folded proteins
(C) hydrophobic amino acid side chains are usually found within the center of protein molecules
(D) intrachain disulfide bonds are usually extremely important in determining the folding of a newly synthesized protein
(E) primary structure is one of the most important factors determining the higher order structure of proteins

90. All the following statements correctly describe γ-immunoglobulins EXCEPT

(A) polypeptide chains composing immunoglobulins are held together by covalent bonds
(B) each immunoglobulin has one antigen-binding site per molecule
(C) heavy immunoglobulin chains have constant C-terminal regions
(D) light immunoglobulin chains have variable N-terminal regions
(E) antigen-binding sites of all different antibodies are determined prior to encountering specific antigens

91. The symptoms of the disease homocystinuria (high levels of homocystine in the urine) may be all the following EXCEPT

(A) treated with dietary supplements of vitamin B_6
(B) treated with dietary supplements of cysteine
(C) the cause of high homoserine levels
(D) a result of the deficiency of cystathionine synthetase
(E) caused by the inability to form methionine from homocysteine by methylation

92. A decreased affinity of hemoglobin for O_2 may result from all the following EXCEPT

(A) initial binding of O_2 to one of the four sites available in each deoxyhemoglobin molecule
(B) high pH
(C) high CO_2 levels
(D) high 2,3-diphosphoglycerate (DPG) levels within erythrocytes
(E) acidosis

93. All the following are amino acids common to the urea cycle EXCEPT

(A) aspartate
(B) citrulline
(C) arginine
(D) ornithine
(E) fumarate

DIRECTIONS: Each group of questions below consists of lettered headings followed by a set of numbered items. For each numbered item select the **one** lettered heading with which it is **most** closely associated. Each lettered heading may be used **once, more than once, or not at all.**

Questions 94–95

For each description below, select the bond that is the most appropriate match.

(A) Ionic
(B) Disulfide
(C) Peptide
(D) Hydrophobic
(E) Hydrogen

94. A bond that contributes to forming the secondary structure of proteins but not the primary structure

95. A covalent bond that can be involved in forming the tertiary structure of proteins once they are synthesized

Questions 96–97

For each familial disorder listed below, select the compound that accumulates in the urine.

(A) Tryptophan
(B) Homocysteine
(C) Dopa
(D) Phenylpyruvate
(E) α-Keto acids

96. Phenylketonuria (PKU)

97. Maple syrup urine disease

Questions 98–99

Match each parameter with the terms of measure used to describe it.

(A) Millimoles/liter
(B) Units/milligram protein
(C) Micromoles/minute
(D) Units/minute
(E) K1

98. K_m (Michaelis constant)

99. Specific activity of enzyme

Questions 100–103

For each description below, select the amino acid that is the most appropriate match.

(A) Cystine
(B) Arginine
(C) Tyrosine
(D) Glutamine
(E) Proline
(F) Serine
(G) Leucine

100. Uncharged derivative of an acidic amino acid

101. Hydroxylation following incorporation into a protein

102. Hydrophobic nature and position in the interior of proteins

103. Basic amino acid with a side-chain pK of approximately pH 12

Questions 104–107

For each description below, select the answer that is the most appropriate match.

(A) Bilirubin diglucuronide
(B) Stercobilin
(C) Biliverdin
(D) Urobilinogen
(E) Heme
(F) Bilirubin
(G) Urobilin

104. A major component of bile

105. Production of brown color of stools

106. Formation by cleavage of heme

107. A major component of urine

Questions 108–109

For each anatomic entity select the substance of which it is primarily composed.

(A) Collagen
(B) Troponin
(C) Keratin
(D) Fibrin
(E) Proteoglycan

108. Ground substance

109. Tendon

Questions 110–111

For each protease below, select the most appropriate class of enzyme.

(A) Chymotrypsin
(B) Pancreozymin
(C) Papain
(D) Carboxypeptidase A
(E) Pepsin

110. Thiol protease

111. Serine protease

Questions 112–113

In the following enzyme-catalyzed reaction,

$$S + E \underset{k_2}{\overset{k_1}{\rightleftharpoons}} S{\cdot}E \overset{k_3}{\Rightarrow} P + E$$

E, S, and P are the enzyme, substrate, and product, respectively; k_1, k_2, and k_3 are rate constants. For each kinetic parameter that follows, select the expression that best defines it.

(A) $k_3 E_t$ (E_t is the total enzyme concentration)
(B) k_2/k_3
(C) k_2/k_1
(D) $(k_2 + k_3)/k_1$
(E) None of the above

112. V_{max}

113. K_m (Michaelis constant)

Amino Acids, Proteins, and Enzymes

Answers

1. The answer is A. *(Stryer, 4/e, pp 18–23.)* Most amino acids in solution at neutral pH are dipolar in charge, so-called zwitterions. The amino group is protonated (NH_3^+) and the carboxyl group is ionized (COO^-). Most amino acids are dipolar at neutral pH owing to the p*K* values of the amino and carboxyl groups. The pH at which amino and carboxyl groups are half dissociated (p*K* values) is about pH 9 and pH 2, respectively. Thus, at neutral pH most amino acids contain both a positive and a negative charge. Molecules that act as both proton donors and proton acceptors are called *ampholytes.*

2. The answer is B. *(Rawn, pp 54–56. Stryer, 4/e, pp 20–21, 647–650.)* Tyrosine is one of three amino acids containing aromatic side chains. It is virtually identical to phenylalanine, except that its phenol contains a hydroxyl group. In contrast to tyrosine, phenylalanine and tryptophan are highly hydrophobic. The presence of the reactive hydroxyl group makes tyrosine much less hydrophobic. The aromatic rings of tyrosine, phenylalanine, and tryptophan interact with other aromatic ring amino acids by pi-electron overlaps.

3. The answer is E. *(Stryer, 4/e, pp 157–160.)* Oxygen binds cooperatively to hemoglobin. That is why the plot of the saturation of hemoglobin in the presence of increasing oxygen is a sigmoidal curve. As each molecule of oxygen is bound to hemoglobin, the strength of the oxygen bound is increased. The cooperative binding by hemoglobin makes hemoglobin a more effective oxygen transporter. In contrast, the high concentration of hydrogen ions and CO_2 in the capillaries of tissues promotes the release of oxygen so that CO_2 can be taken up for transport to the lungs. This promotion of the release of O_2 with increased pH and higher levels of CO_2 is called the *Bohr effect.* The high concentration of oxygen in the lungs and the cooperative binding of oxygen leads to the unloading of CO_2 in the lungs, while just the opposite effect occurs in active tissues.

4. The answer is B. *(Stryer, 4/e, pp 447–448, 577–578, 630.)* The comatose emergency room patient is not in an alcoholic stupor because his blood alcohol is below the level of approximately 20 mg/dL where alcohol-related symptoms might be seen. In fact, the legal levels of drunkenness in most states are around 80 to 100 mg/dL. Likewise, the patient is neither hyper- nor hypoglycemic because the blood glucose level is normal at approximately 80 mg/dL. Since the blood alanine aminotransferase (ALT) level is normal, it can be presumed that liver cirrhosis, which is common in alcoholics, is not the cause of the problem. High ALT levels in the blood indicate a release of the enzyme due to liver cell damage. The observation that creatine kinase (CK2) levels and lactate dehydrogenase (LDH) levels are higher than normal would indicate that a myocardial infarction has occurred sometime in the past 9 to 36 h. CK2 and LDH are released from damaged cardiac muscle cells following a heart attack. CK2 activity reaches a peak at about 18 h after the infarction, while LDH activity peaks in the plasma about 36 to 40 h after the infarction. Since both levels are high, it is highly likely that the infarction occurred sometime between 18 and 36 h prior to the blood measurement. The use of plasma enzymes in clinical diagnosis of cancers and other diseases such as heart disease and liver cirrhosis is quite common.

5. The answer is B. *(Stryer, 4/e, pp 237–238.)* Enzymes may be regulated in an acute or chronic manner. Acute or minute-to-minute regulation of enzymes occurs through end-product inhibition, allosteric control, substrate availability, and covalent modification. In this type of regulation, product, substrates, end products, or another regulatory enzyme will change the V_{max} or the K_m of the enzyme in an immediate manner. In contrast, long-term changes in enzyme activity measured in terms of hours or days are best effected by an increase in the degradation rate of the enzyme or the rate of synthesis of the enzyme. This change in the amount of enzyme present usually occurs in response to the presence of hormones or a metabolite such as a repressor or a derepressor. By a variety of acute and chronic mechanisms of control, enzyme levels or activity can be monitored for the long or the short term.

6. The answer is E. *(Stryer, 4/e, pp 53–58.)* By specifically splitting proteins into different, small, overlapping fragments that can be sequenced, the proper placement of each fragment can lead to a determination of the complete amino acid sequence of a protein. Different overlapping fragments can be generated by using more than one specific cleaving agent. In the example shown, the plant peptide isolated by the medical student is first treated with trypsin. Trypsin enzymatically cleaves peptide bonds on the carbonyl side of either arginine or lysine. Since three peptides were produced and subsequently sequenced by Edman degradation, we can see that two sites specific to trypsin cleavage were present on the original peptide. Of the three tryptic peptides, two have trypsin-specific amino acids: lysine and arginine. Chemical cleavage of the unknown peptide with cyanogen bromide yielded two peptides. Cyanogen bromide is specific for the carbonyl side of the methionine residues. Once sequenced, the different peptides produced by cyanogen bromide cleavage can be compared with the peptide produced by the enzymatic cleavage of trypsin. The cyanogen bromide cleavage produces two peptides that overlap the ends of the peptides produced by trypsin. This allows a deduction of the correct ordering of the peptides and, hence, a determination of the sequence of the original plant peptide. The N-terminal and C-terminal ends of the peptides can either be determined by the Edman degradation sequence or by the knowledge that trypsin cleavage produces lysine or arginine with exposed carboxyl ends and cyanogen bromide produces methionine with an exposed carboxyl end. The peptides in the problem shown were written in the conventional manner, starting with the amino-terminal residue.

7. The answer is E. *(Stryer, 4/e, pp 154–157.)* Adult hemoglobin, or hemoglobin A, is composed of four polypeptide chains. Two of the chains are alpha chains and two of the chains are beta chains. The chains are held together by noncovalent interactions. The hemoglobin tetramer can best be represented as being composed of two dimers, each containing the two different polypeptides. Thus the designation $(\alpha_1\text{-}\beta_1)$ $(\alpha_2\text{-}\beta_2)$, which refers to dimers 1 and 2 respectively, is the most correct way to refer to the quaternary structure of adult hemoglobin. Hydrophobic interactions are thought to be the main noncovalent interactions holding all four polypeptides together.

8. The answer is B. *(Stryer, 4/e, pp 28–37.)* The α-helical segments of proteins are one of the most common secondary structures of proteins. The helical structure is composed of a spiraled polypeptide backbone core with the side chains of component amino acids extending outward from the central axis in order to avoid interfering sterically or electrostatically with each other. All the peptide bond carbonyl oxygens are hydrogen-bonded to a peptide linkage that is four residues ahead in the polypeptide. This leads to 3.6 amino acids per turn, spatially held together in the α-helical structure. Since the configuration of the α-helix is compatible with being in the interior of proteins, amino acids with nonpolar, hydrophobic side chains predominate. Conversely, amino acids that are charged or have bulky side chains may interfere with the α-helical structure if present in large enough amounts. Proline and hydroxyproline are not at all compatible with the right-handed spiral of the α-helix. They insert a kink in the chain. Likewise, large numbers of charged amino acids such as lysine or histidine disrupt the helix by forming electrostatic bonds or by ionically repelling one another. In addition, amino acids with bulky side chains such as tryptophan or isoleucine will also tend to disrupt the configuration of the α-helix.

9. The answer is D. *(Stryer, 4/e, pp 577–578.)* Isozymes are multiple forms of a given enzyme that occur within a given species. Since isozymes are composed of different proteins, analysis by electrophoretic separation can be done. Lactate dehydrogenase is a tetramer composed of any combination of two different polypeptides, H and M. Thus, the possible combinations are H4, H3M1, H2M2, H1M3, and M4. Although each combination is found in most tissues, M4 predominates in the liver and skeletal muscle while H4 is the predominant form in the heart. White and red blood cells as well as brain cells contain primarily intermediate forms. The M4 forms of the isozyme seem to have a higher affinity for pyruvate compared with the H4 form. Following a myocardial infarction, the H type of lactate dehydrogenase rises and reaches a peak approximately 36 h later. Plasma levels of the H type of lactate dehydrogenase are an indicator used to assess myocardial infarctions.

10. The answer is D. *(Stryer, 4/e, pp 28–37.)* The spiral structures of α-helices are held together by hydrogen bonds between the carbonyl oxygens and the amide nitrogens. Amino acids with uncharged or hydrophobic side chains are most compatible with formation of α-helical structures. Consequently, a polypeptide composed of uncharged amino acids or hydrophobic amino acids tends to be able to form and maintain an α-helix. Therefore, the structure shown in answer A contains only amino acids with hydrophobic side chains and, thus, is relatively unaffected by pH. In contrast, the structures shown in answers B and E would only be uncharged at extremely low pH (below pH 4), since they are polyacidic amino acids composed of either aspartate or glutamate. At very low pH, the carboxyl group on the side chain of these amino acids will tend to be proteinated and α-helical structures can be formed. However, above pH 4, the carboxyl groups contain negative charges, which electrostatically repulse one another and destroy the α-helical spiral. Likewise, the polypeptide shown in answer C contains numerous lysines that are positively charged until a very high pH (approximately pH 12) is reached. Thus, the repetitive positive charges tend to disrupt an α-helical structure. We are then left with answer D, Leu-His-Gly-His-Ala-His-Leu-Leu, which is a structure composed of hydrophobic amino acids and repeating units of the amino acid histidine. Below the p*K* of pH 6, histidine will be relatively uncharged and thus capable of forming an α-helix. As the pH is raised, histidine gradually acquires a positive charge, which tends to disrupt the helical structure and produce an extended conformation. Thus, only this repeating polypeptide is capable of transforming from an α-helical structure to an extended structure when going from pH 4 to pH 8.

11. The answer is D. *(Stryer, 4/e, pp 45–53, 60–63.)* SDS-polyacrylamide gel electrophoresis is used to separate proteins based upon their mass. In this process, proteins are dissolved in a solution of the detergent sodium dodecyl sulfate (SDS). SDS is an ionic detergent that disrupts noncovalent interactions in native proteins and binds to the proteins at a ratio of about 1 SDS molecule per 2 amino acid residues. Thus, the denatured protein ends up with a net negative charge that is approximately proportional to the mass of the protein. Consequently, the original charge of the proteins plays no role in the electrophoresis pattern. Since mass is dependent on the quantity of matter in a given space, upon application of direct electrical current, small proteins move quickly through the gel while larger proteins stay near the point of application. In contrast, isoelectric focusing is dependent upon the native charge of the proteins themselves. In this technique, which is dependent upon the relative content of acidic and basic residues of each given protein, mixtures of proteins are electrophoresed in a pH gradient in a gel. Each protein moves until it reaches a position of gel at which the pH is equal to the pI of the protein. Thus, the method of separation is dependent upon the isoelectric point of the protein. In dialysis, semipermeable membranes (which allow small but not large molecules through) separate groups of proteins by their size class. Size, which is the amount of space occupied by a given protein, is thus distinct from mass. Charge is also used to separate proteins using ion exchange chromatography. For example, if proteins are passed over a column of beads containing negative charges (e.g., carboxylate groups), then proteins with a net positive charge will bind to the column of beads, whereas a net negatively charged protein will not. In this manner, positively charged proteins bound to such columns can then later be eluted by increasing the concentration of sodium chloride or some other salt in the column solution. The converse is true using resin or columns of beads containing positive charges. The most specific method for separating out proteins is western blotting or some variation thereof. In this technique, samples of proteins that have been electrophoresed on an SDS-polyacrylamide gel are transferred (i.e., blotted) to a nitrocellulose sheet to make them more assessable for reaction. Subsequently, an antibody specific for the protein of interest is added to the sheet. The antibody-antigen complex on the sheet can then be detected by rinsing the sheet with a second, radioactively labeled antibody that is specific for the primary antibody. The radioactivity of the second antibody produces a dark band on x-ray film called an *autoradiogram.* Another identification method is to use an enzyme on the second antibody to generate a colorimetric product as in the enzyme-linked immunosorbent assay (ELISA) method.

12. The answer is E. *(Stryer, 4/e, pp 31–32.)* Collagen has an unusual amino acid composition in that approximately one-third of the collagen molecules are glycine. The amino acid proline is also present in a much greater amount than in other proteins. In addition, two somewhat unusual amino acids are found in collagen, 4-hydroxyproline and 5-hydroxylysine. Hydroxyproline and hydroxylysine, per se, are not incorporated during the synthesis of collagen. Proline and lysine are hydroxylated by specific hydroxylases after collagen is synthesized. A reducing agent such as ascorbate (vitamin C) is needed for the hydroxylation reaction to occur. In its absence, the disease known as scurvy occurs. Only proline or lysine residues located on the amino side of glycine residues are hydroxylated. The hydroxylysine residues of collagen are important sites of glycosylation of disaccharides of glucose and galactose.

13. The answer is B. *(Stryer, 4/e, pp 18–23.)* Proteins can be effective buffers of body and intracellular fluids. Buffering capacity is dependent upon the presence of amino acids having ionizable side chains with p*K*'s near physiologic pH. In the example given, only histidine has an ionizable imidazolium group that has a p*K* close to neutrality (p*K* = 6.0). Valine and leucine are amino acids with uncharged, branched side chains. Lysine has a very basic amino group (p*K* = 10.5) on its aliphatic side chain that is positively charged at physiologic pH, and aspartic acid has a side chain carboxyl (p*K* = 3.8) that is negative at pH 7.

14. The answer is A. *(Stryer, 4/e, pp 630–631, 636.)* Oxaloacetate has many possible fates. It can be converted into pyruvate, citrate, or phosphoenolpyruvate or be transaminated with glutamate to form aspartate. In the reverse reaction, another intermediate of the citric acid cycle, α-ketoglutarate, is transaminated with aspartate to form oxaloacetate and glutamate in the presence of the enzyme aspartate aminotransferase:

$$\text{aspartate} + \alpha\text{-ketoglutarate} \leftrightarrow \text{oxaloacetate} + \text{glutamate}$$

Alanine is the transamination product of pyruvate, which is not a member of the citric acid cycle. Serine, lysine, and valine are not formed by transamination in humans.

15. The answer is A. *(Stryer, 4/e, pp 18–23.)* At neutral pH, amino acids in solution are zwitterions (i.e., dipolar ions) containing both a protonated amino group and a dissociated carboxyl group. The charged groups of amino acids are shown below in relation to acid, neutral, and basic pH:

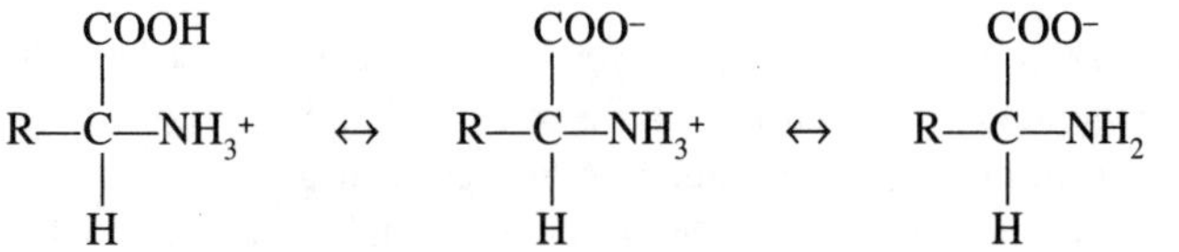

$$\underset{\text{Acid (pH 1)}}{R\text{—}\overset{\overset{\text{COOH}}{|}}{\underset{\underset{\text{H}}{|}}{C}}\text{—}NH_3^+} \leftrightarrow \underset{\text{Neutral (pH 7)}}{R\text{—}\overset{\overset{\text{COO}^-}{|}}{\underset{\underset{\text{H}}{|}}{C}}\text{—}NH_3^+} \leftrightarrow \underset{\text{Basic (pH 14)}}{R\text{—}\overset{\overset{\text{COO}^-}{|}}{\underset{\underset{\text{H}}{|}}{C}}\text{—}NH_2}$$

Since some amino acids do have polar side chains (R groups), these amino acids will also contain other charges in addition to the charged carboxyl and amino groups that are common to all amino acids.

16. The answer is B. *(Rawn, pp 39–41. Stryer, 4/e, pp 42–43.)* The acidity of aqueous solutions is expressed as the pH and is equal to the negative logarithm to the base 10 of the concentration of hydrogen ions ($[H^+]$). Thus for pure water, which has an $[H^+]$ concentration of 1.0×10^{-7} molar, pH = $-\log 10^{-7} = -(-7) = 7.0$.

17. The answer is E. *(Stryer, 4/e, pp 237–238.)* Activation or inactivation of certain key regulatory enzymes is accomplished by covalent modification of specific serine or tyrosine hydroxyl residues by phosphorylation. For example, skeletal muscle glycogen phosphorylase *b* is activated by phosphorylation of a single serine residue (serine 14) in each subunit of the dimers composing the enzyme. The phosphorylation reaction itself is catalyzed by phosphorylase kinase. In turn, activated muscle glycogen phosphorylase *a* is deactivated by a specific phosphatase that hydrolyzes the phosphoryl group off of serine 14.

18. The answer is C. *(Stryer, 4/e, pp 719–721.)* Serine is transformed to glycine upon contributing one carbon atom to tetrahydrofolate to form 5,10-methylene-tetrahydrofolate. This is an important contribution to the one-carbon pool of tetrahydrofolate derivatives, which serve as carbon donors in a number of biosynthetic reactions. Serine can be synthesized de novo from intermediates of the glycolytic pathway. Histidine and glycine are other important sources of one-carbon units.

19. The answer is D. *(Stryer, 4/e, pp 28–30.)* Keratins are a type of intermediate filament that comprises a large portion of many epithelial cells. The characteristics of skin, nails, and hair are all due to keratins. Keratins contain a large amount of the disulfide amino acid cystine. Approximately 14 percent of the protein composing human hair is cystine. This is the chemical basis of depilatory creams, which are reducing agents that render keratins soluble by breaking the disulfide bridges of these insoluble proteins. The basic structure of intermediate filament proteins is a two- or three-stranded α-helical core of 300 amino acids in length.

20. The answer is C. *(Stryer, 4/e, pp 53–56.)* Edman degradation removes one residue at a time from the amino end of a peptide. By reacting phenyl isothiocyanate with the uncharged amino terminal group, a phenylthiocarbamoyl derivative is formed. Mild acidic conditions cause release of the cyclic derivative of the terminal amino acid without destruction of the rest of the peptide. Thus, the phenylthiohydantoin amino acid (PTH-amino acid) can be identified by chromatography, and the intact peptide shortened by one amino acid can undergo another round of Edman degradation. By analysis of specific peptides cleaved from a larger protein, the entire amino acid sequence of a protein can be worked out. The Edman process has been incorporated into the design of the sequenator, an instrument for the automatic determination of protein structure.

21. The answer is E. *(Stryer, 4/e, pp 185–188.)* Enzymes accelerate reactions by decreasing the Gibbs free energy of activation or the activation barrier. By combining with the substrate, an enzyme alters the reaction pathway to one whose transition state energy is lower than the original, uncatalyzed pathway. In this manner, reactions are accelerated by enzymes by millions of times. The energy levels of reactants are not affected by enzymes. Enzymes never change reaction equilibria.

22. The answer is B. *(Stryer, 4/e, pp 169–171.)* A single amino acid mutation is responsible for sickle cell. Valine is substituted for glutamate at position 6 (from the amino acid terminal) of the beta chain. Since valine has a nonpolar side chain, while glutamate has a charged carboxyl group on its side chain at pH 8.6, electrophoresis reveals that hemoglobin S has a greater overall positive charge than does normal hemoglobin (hemoglobin A). In patients heterozygous for the sickle cell gene (called *sickle cell trait*), both hemoglobin S and hemoglobin A will be produced. Thus, two bands will be observed on electrophoretic gels.

23. The answer is A. *(Rawn, pp 58–60. Stryer, 4/e, pp 42–43.)* The isoelectric point (pI) of an amino acid is that pH at which the net charge is zero. Since pK values denote the pH at which a given α-COOH, α-NH_3^+, or R group is dissociated, it is possible to calculate the pI. For amino acids with uncharged side groups the pI is simply the halfway point between the α-COOH pK1 and the α-NH_3^+ pK2: (pK1 + pK2)/2. For basic amino acids it is the average of the α-NH_3^+ and the side chain group. If the side chain group is designated as pK3, then pI = (pK2 + pK3)/2. For acidic amino acids the pI is halfway between the α-COOH and the side chain group: (pK1 + pK3)/2. For aspartate, $(2.0 + 3.9)/2 = 3.0$.

24. The answer is A. *(Stryer, 4/e, pp 46–48.)* Proteins to be separated by electrophoresis on polyacrylamide gels are dissolved in sodium dodecyl sulfate (SDS) and mercaptoethanol or dithiothreitol. The anionic detergent (SDS) disrupts all noncovalent interactions in the proteins, binding to the protein chains at about a ratio of one SDS to two amino acid residues. Thus, the native charge of the protein is rendered insignificant and the protein gains a large net negative charge. Mercaptoethanol or dithiothreitol reduces disulfide bonds, which further separates protein subunits. The denatured protein migrates in a negative-to-positive direction.

25. The answer is B. *(Stryer, 4/e, pp 647–650.)* Lack of the enzyme homogentisate oxidase results in alkaptonuria, an autosomal recessive disease. Homogentisate, a metabolite in the pathway of degradation of phenylalanine and tyrosine, accumulates. It is excreted in the urine, where it oxidizes and is polymerized to a dark substance upon standing. Under normal conditions, phenylalanine is degraded to tyrosine, which is broken down through a series of steps to fumarate and acetoacetate.

26. The answer is E. *(Stryer, 4/e, pp 31–33.)* Hydroxyproline and hydroxylysine are not present in newly synthesized collagen. Proline and lysine residues are modified by hydroxylation in a reaction requiring the reducing agent ascorbic acid (vitamin C).

$$\underset{\text{residue}}{\text{prolyl}} + O_2 + \alpha\text{-ketoglutarate} \rightarrow \underset{\text{residue}}{\text{hydroxyprolyl}} + CO_2 + \text{succinate}$$

The enzymes catalyzing the reactions are the dioxygenases prolyl hydroxylase and lysyl hydroxylase. In scurvy, which results from a deficiency of vitamin C, insufficient hydroxylation of collagen causes malformed connective tissue fibers, which are manifest as degeneration of blood vessels and loss of skin attachment.

27. The answer is A. *(Stryer, 4/e, pp 634–636.)* In humans, excess NH_4^+ is converted to urea for excretion by the kidneys. Initially, CO_2 and NH_4^+ condense to form carbamoyl phosphate. Ornithine then combines with carbamoyl phosphate to produce citrulline in the first step of the urea cycle proper. In the second step, argininosuccinate synthetase catalyzes the condensation of citrulline and aspartate to yield argininosuccinate. This reaction requires adenosine triphosphate (ATP), which is broken down to adenosine monophosphate (AMP) and pyrophosphate (PP_1). Argininosuccinate is cleaved into arginine and fumarate. Finally, arginine is hydrolyzed to urea and ornithine.

28. The answer is E. *(Rawn, pp 75–81. Stryer, 4/e, pp 27–37.)* In the illustration accompanying the question, Figure A is a random coil; Figure B an α-helix; Figure C a pleated sheet; and Figure D a triple-stranded helix. Figure E is a super-coiled helix, one of the forms of hair keratin.

29. The answer is E. *(Stryer, 4/e, pp 31–32.)* Tropocollagen has the structure of a triple helix because many proline and hydroxyproline residues prevent the hydrogen bond formation necessary for an α-helix (Figure B in the illustration accompanying the questions). It is the basic unit of collagen fibrils obtained through extraction of insoluble collagen with dilute acid.

30. The answer is B. *(Stryer, 4/e, pp 402–404.)* Calcium ions are the regulators of contraction of skeletal muscle. Calcium is actively sequestered in sarcoplasmic reticulum by an ATP pump during relaxation of muscle. Nervous stimulation leads to the release of calcium into the cytosol and raises the concentration from less than 1 μM to about 10 μM. The calcium binds to troponin C. The calcium-troponin complex undergoes a conformational change, which is transmitted to tropomyosin and causes tropomyosin to shift position. The shift of tropomyosin allows actin to interact with myosin and contraction to proceed.

31. The answer is E. *(Stryer, 4/e, pp 24–25.)* The structure shown has been drawn in an unconventional manner. Convention dictates that peptides be drawn with the *N* terminal to the left and the *C* terminal to the right. Likewise, peptides are always named from the *N*-terminal end. The tripeptide shown is threonine-asparagine-serine.

32. The answer is D. *(Stryer, 4/e, pp 53–58.)* Acid hydrolysis of all the peptides listed in the question would produce equivalent amounts of the three given amino acids. Trypsin is a pancreatic endopeptidase that cleaves the peptide bond at the carboxyl end of the basic amino acids arginine and lysine. Only the trypsinization of Ala-Lys-Gly will yield free glycine and a single dipeptide.

33. The answer is C. *(Stryer, 4/e, pp 27–28.)* The peptide bond of proteins is the bond between the carboxyl carbon of one amino acid residue and the amino nitrogen of the adjacent amino acid residue.

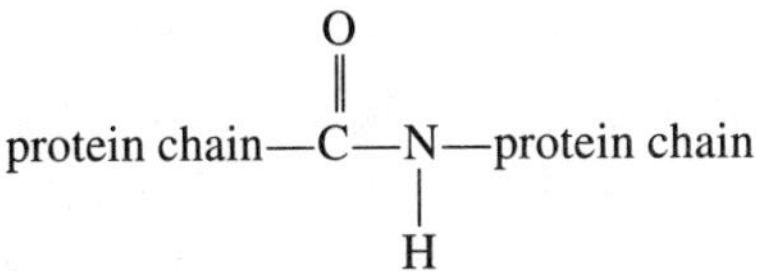

```
                 O
                 ‖
protein chain—C—N—protein chain
                    |
                    H
```

It is rigid and planar, having no rotational freedom, such that the carbonyl oxygen and the amino hydrogen are always in a *trans* (opposite) position.

However, because of this conformation, adjacent bonds to the α-carbon of the amino acid residues on either side of the peptide bond have a large degree of freedom of rotation.

34. The answer is A. *(Stryer, 4/e, pp 735–736.)* Stercobilin is the pigment that imparts the characteristic color to stools. Bilirubin diglucuronide is hydrolyzed to yield bilirubin, which is reduced to urobilinogen by bacterial flora. Urobilinogen is further reduced to stercobilinogen. Oxidation of stercobilinogen in air yields stercobilin.

35. The answer is C. *(Rawn, p 184. Stryer, 4/e, pp 192–195.)* In the Lineweaver-Burk transformation of the Michaelis-Menten equation for the rate of enzymatic processes, the reciprocal of velocity may be plotted on the *y* axis against the reciprocal of substrate concentration on the *x* axis. Determination of K_m and V_{max} is made possible by direct graphic measurement of both slope and *x*-axis intercept.

36. The answer is D. *(Stryer, 4/e, pp 577–578.)* Different forms of the same enzyme, isozymes, are formed from the assembly of different subunits in varying proportions. For example, lactate dehydrogenase is a tetramer of four subunits. The four subunits can all be the same or be a varying proportion of two kinds of subunits (H and M). Combinations of the two subunits can result in five different isozymes: H_4, H_3M, H_2M_2, HM_3, M_4. These enzymes have similar substrate specificities but different affinities and electrophoretic mobilities. The H type is found mainly in heart muscle, while the M type is found in great abundance in liver and skeletal muscle.

37. The answer is A. *(Stryer, 4/e, pp 197–198.)* Competitive inhibition kinetics are defined as effects that increase K_m of an enzyme without affecting V_{max}. Thus the double-reciprocal plot changes in slope but not in *y*-axis intercept as enzyme concentration is altered. Competitive inhibition can be overcome at a sufficiently high concentration of substrate. In contrast, noncompetitive inhibition affects the V_{max}, not the K_m, and is irreversible.

38. The answer is E. *(Stryer, 4/e, pp 634–636.)* All the compounds listed are intermediates of the citric acid cycle. However, only fumarate is an intermediate of both the citric acid and urea cycles. It and arginine are produced from argininosuccinate. Once produced by the urea cycle, fumarate enters the citric acid cycle and is converted to malate and then oxidized to oxaloacetate. Depending upon the organism's needs, oxaloacetate can either enter gluconeogenesis or react with acetyl CoA to form citrate.

39. The answer is D. *(Rawn, pp 34–35. Stryer, 4/e, pp 186–187.)* In the problem presented in the question, assuming that

$$K_{eq} = \frac{(C)}{(A)(B)} = 0.1$$

then

$$\Delta G° = -2.3\ RT \log K_{eq} = -2.3\ RT \log (0.1) = 2.3\ RT.$$

40. The answer is E. *(Rawn, pp 166–173. Stryer, 4/e, pp 192–195.)* In the question presented, the intercept of the double-reciprocal plot on the x-axis is $-1/K_m$. Since $-1/K_m = -3$, $K_m = 0.33$. Similarly, from the observed y-axis intercept ($1/V_{max}$), the slope may be determined.

41. The answer is B. *(Stryer, 4/e, pp 193–195.)* When an enzyme obeys classic Michaelis-Menten kinetics as seen in the figure, then the Michaelis constant (K_m) and the maximal rate (V_{max}) can be readily derived. By plotting a reciprocal of the Michaelis-Menten equation, a straight-line, Lineweaver-Burk plot is produced. The y intercept is $1/V_{max}$, while the x intercept is $-1/K_m$. Thus, a reciprocal of these absolute values yields V_{max} and K_m.

42. The answer is C. *(Stryer, 4/e, pp 198–199.)* Allosteric enzymes, unlike simpler enzymes, do not obey Michaelis-Menten kinetics. Often, one active site of an allosteric enzyme molecule can positively affect another active site in the same molecule. This leads to cooperativity and sigmoidal enzyme kinetics in a plot of $[S]$versus V. The terms *competitive* and *noncompetitive inhibition* apply to Michaelis-Menten kinetics and not allosteric enzymes.

43. The answer is A. *(Stryer, 4/e, pp 192–195.)* As shown on a double-reciprocal plot of enzyme activity rate versus substrate concentration, a competitive inhibitor that increased K_m without affecting V_{max} would increase the slope without affecting the intercept on the y axis. On the graph presented in the question, such competitive inhibition would be represented by curve A, where curve X is the condition of no inhibition.

44. The answer is D. *(Stryer, 4/e, pp 368–369, 375–376.)* Immunoglobulin G is composed of pairs of light chains and heavy chains attached by disulfide bridges. If the reducing agent mercaptoethanol is used to break the disulfide bridges and urea is used to disrupt noncovalent interactions, two identical light subunits (25 kd) and two identical heavy chains (50 kd) per protein can be resolved with electrophoresis. A small amount of carbohydrate is also present. In contrast, the proteolytic enzyme papain cleaves the heavy chains, which results in two Fab molecules consisting of the entire light chain attached to the amino terminal half of each heavy chain and two Fc molecules consisting of the carboxyl terminal half of each heavy chain. Other proteolytic enzymes are nonspecific. Levels rise and fall in the serum dependent upon specific induction by antigen.

45. The answer is B. *(Stryer, 4/e, pp 48–50.)* Each of the techniques listed separates proteins from each other and from other biologic molecules based upon characteristics such as size, solubility, and charge. However, only affinity chromatography can use the high affinity of proteins for specific chemical groups or the specificity of immobilized antibodies for unique proteins. In affinity chromatography a specific compound that binds to the desired protein—such as an antibody, a polypeptide receptor, or a substrate—is covalently bound to the column material. A mixture of proteins is added to the column under conditions ideal for binding the protein desired, and the column is then washed with buffer to remove unbound proteins. The protein is eluted by either adding a high concentration of the original binding material or by making the conditions unfavorable for binding (e.g., changing the pH). The other techniques are less specific than affinity binding for isolating proteins. Dialysis separates large proteins from small molecules. Ion-exchange chromatography separates proteins with an overall charge of one sort from those proteins with an opposite charge (e.g., negative from positive). Gel-filtration chromatography separates on the basis of size. Electrophoresis separates proteins on the principle that net charge influences the rate of migration in an electric field.

46. The answer is C. *(Stryer, 4/e, pp 634–636.)* The steps of the urea cycle are divided between the mitochondrial matrix and cytosol of liver cells in mammals. The formation of ammonia, its reaction with carbon dioxide to produce carbamoyl phosphate, and the conversion to citrulline occur in the matrix of mitochondria. Citrulline diffuses out of the mitochondria, and the next three steps of the cycle, which result in the formation of urea, all take place in the cytosol.

47. The answer is E. *(Stryer, 4/e, pp 636–637.)* Partial blockage of the urea cycle leads to conditions ranging from lethargy and episodic vomiting to mental retardation. A complete block is incompatible with life. A major reason for the toxicity is the severe depletion of ATP levels caused by the siphoning off of α-ketoglutarate from the citric acid cycle in an attempt to consume ammonia. Glutamate dehydrogenase and glutamine synthetase, respectively, catalyze the following reaction:

α-ketoglutarate + NH_4^+ → glutamate + NH_4^+ → glutamine

As can be seen, this is the reverse order of steps whereby glutamine is successively deaminated to first glutamate and then α-ketoglutarate by the enzymes glutaminase and glutamate dehydrogenase, respectively. It is thought that the high level of ammonia ions shifts the equilibrium of the dehydrogenase in favor of the formation of glutamate.

48. The answer is B. *(Stryer, 4/e, pp 23, 42–43.)* To reach pH 7.0, approximately 100 percent of the α-carboxyl group (pK_{a1} = 2.19) and 90 percent of the γ-carboxyl group (pK_{a2} = 4.25) of glutamic acid must be dissociated. At that pH, approximately twice the amount of NaOH as there are glutamic acid molecules has been utilized to titrate the two carboxyl groups. Since each milliliter of a 1 *M* NaOH solution contains 1 mmol of OH^-, about 3 mmol of the amino acid is present.

$$\underset{\text{pH 1.0}}{HOOC(CH_2)_2CH(NH_3^+)COOH} + 2OH^- \rightarrow \underset{\text{pH 7.0}}{^-OOCCH_2CH(NH_3^+)COO^-} + 2H_2O$$

49. The answer is B. *(Stryer, 4/e, pp 23, 42–43.)* At its isoelectric pH, an amino acid has a net charge of zero. Since aspartic acid has three dissociable groups, the pH_I is the average of the two pK_a values on either side of the isoelectric species.

$$pH_I = \frac{pK_{a1} + pK_{a2}}{2}$$

$$\underset{+1}{COOH,COOH,NH_3^+} \underset{2.09}{\overset{pK_{a1}}{\rightleftharpoons}} \underset{0}{COOH,COO^-,NH_3^+} \underset{3.86}{\overset{pK_{a2}}{\rightleftharpoons}} \underset{-1}{COO^-,COO^-,NH_3^+} \underset{9.82}{\overset{pK_{a3}}{\rightleftharpoons}} \underset{-2}{COO^-,COO^-,NH_2}$$

50. The answer is B. *(Stryer, 4/e, pp 730, 1002.)* Thyroxine is a derivative of tyrosine. It is formed by the iodination and joining of peptide-linked tyrosyl residues of thyroglobulin. Proteolysis of thyroglobulin yields thyroxine. Thyroxine is also called *tetraiodothyronine,* or T_4, because of the four iodine atoms of the thyroid hormone.

51. The answer is D. *(Rawn, pp 131–141. Stryer, 4/e, pp 157–168.)* Hemoglobin is a tetrameric hemoprotein whose oxygen saturation curves exhibit sigmoidal kinetics because of cooperative interactions among the four binding sites. Oxygen is bound to hemoglobin without changing the redox state of the iron from the ferrous state. Carbon monoxide and cyanide both bind to hemoglobin more tightly than does oxygen itself. O_2 is released to tissues and exchanged with CO_2 since increased CO_2 levels in capillaries lead to decreased affinity of hemoglobin for O_2.

52. The answer is B. *(Stryer, 4/e, pp 192–195.)* When a modifier binds at the allosteric site, it affects the active site by altering V_{max} and K_m. The substrate binds to the active, or catalytic, site, where it is modified. Binding of both substrate and modifier is, of course, concentration-dependent. The velocity of an allosteric enzyme reaction depends on the concentration of both the substrate and the modifier.

53. The answer is C. *(Stryer, 4/e, pp 629–633.)* Most amino acids are deaminated by transamination, whereby the α-amino group is transferred to either α-ketoglutarate to form glutamate or to pyruvate to form alanine. The carbon skeletons of the remaining α-keto acids are then catabolized further. However, direct deamination of the α-amino groups of serine and threonine to NH_4^- can occur because of the presence of the side-chain hydroxyl groups. Serine dehydratase and threonine dehydratase catalyze a two-step reaction in which dehydration precedes deamination.

$$\text{Threonine} \rightarrow \alpha\text{-ketobutyrate} + H_2O + NH_4^+$$
$$\text{Serine} \rightarrow \text{pyruvate} + H_2O + NH_4^+$$

54. The answer is D. *(Stryer, 4/e, pp 529–531, 544–548.)* There are thought to be three sites of energy conservation in the mitochondrial electron transport chain that lead to the production of ATP. These sites are found between NADH and a flavoprotein (site I), between cytochrome *b* and cytochrome *c* (site II), and between cytochrome *c* and cytochromes *a*, a_3 (site III). At these sites, protons are pumped across the inner mitochondrial membrane as elecrons flow along the respiratory chain from NADH to O_2. The proton-motive force is used to synthesize ATP using the enzyme ATP synthase. This enzyme is also known as *mitochondrial* H^+*-ATPase* because it was initially discovered in broken mitochondrial preparations, where it will hydrolyze ATP.

55. The answer is C. *(Stryer, 4/e, pp 48–50.)* Proteins can be separated on the basis of their overall charge at a given pH by ion exchange chromatography. At low pH all proteins have an overall positive charge since carboxyl groups are protonated. Thus, proteins will tend to bind to a cation-exchange column that has immobilized the negative charges. Usually negatively charged sulfonic polystyrene resin is used, and Na^+ charges are exchanged for the positively charged protein groups. Once binding has occurred, the pH and NaCl concentration of the eluting medium are increased, and proteins that have a low density of net negative charge will emerge first, with those having a higher density of negative charge following. The only information that can be obtained from the information given in the question is that the enzyme has been purified over 100-fold. The turnover rate of the enzyme cannot be deduced. Likewise, the yield, which is the amount of original enzyme protein recovered, cannot be determined. The structure of the enzyme is not revealed by the information given.

56. The answer is E. *(Stryer, 4/e, pp 630–631, 634–636.)* Aspartate is a glucogenic amino acid that is also used to carry NH_4^+ into the urea cycle. Aspartate aminotransferase catalyzes the direct transamination of aspartate to oxaloacetate.

$$\text{Aspartate} + \alpha\text{-ketoglutarate} \leftrightarrow \text{oxaloacetate} + \text{glutamate}$$

Oxaloacetate may either be utilized in the citric acid cycle or undergo gluconeogenesis. Argininosuccinate synthetase catalyzes the condensation of citrulline and aspartate to form argininosuccinate.

$$\text{Citrulline} + \text{asparate} + \text{ATP} \leftrightarrow \text{argininosuccinate} + \text{AMP} + \text{PP}_i$$

In this manner, one of the two nitrogens of urea is introduced into the urea cycle.

57. The answer is B. *(Stryer, 4/e, pp 392–398.)* In skeletal muscle two kinds of interacting protein filaments are found. Thick filaments of 15 nm in diameter contain primarily myosin. Thin filaments of 7 nm in diameter are composed of actin, troponin, and tropomyosin. The thick and thin filaments slide past one another during muscle contraction. Myosin is an ATPase that binds to thin filaments during contraction. α-Actinin can be found in the Z line.

58. The answer is A. *(Stryer, 4/e, pp 198–199, 237–247.)* The binding of an effector to the regulatory subunit of an allosteric enzyme causes a conformational change that either increases or decreases the activity of the enzyme's separate catalytic site. Only in some allosteric molecules, such as hemoglobin, does positive cooperativity occur. A positive effector increases substrate binding. This is the case with cyclic AMP–dependent protein kinase of the glycogen phosphorylase cascade. Cyclic AMP binds the regulatory subunit that dissociates from the catalytic subunit and thereby activates it. In the absence of cyclic AMP, the regulatory subunit tightly binds the catalytic subunit and inactivates the enzymes. Many allosteric enzymes are often placed at the first, or committed, step of a metabolic pathway. The end product of the pathway then acts as a negative effector of the enzyme. This is called *feedback inhibition.* An allosteric enzyme does not obey Michaelis-Menten kinetics.

59. The answer is D. *(Rawn, pp 157–165, 195–201. Stryer, 4/e, pp 250–251.)* Pepsin is secreted in a proenzyme form in the stomach. Unlike the majority of proenzymes, it is not activated by protease hydrolysis. Instead, spontaneous acid hydrolysis at pH 2 or lower converts pepsinogen to pepsin. Hydrochloric acid secreted by the stomach lining creates the acid environment. All the enzymes secreted by the pancreas are activated at the same time upon entrance into the duodenum. This is accomplished by trypsin hydrolysis of the inactive proenzymes trypsinogen, chymotrypsinogen, procarboxypeptidase, and proelastase. Primer amounts of trypsin are derived from trypsinogen by the action of enteropeptidase secreted by the cells of the duodenum.

60. The answer is E. *(Stryer, 4/e, pp 237–238, 247–252.)* Aspartate transcarbamoylase, which controls the rate of pyrimidine synthesis in mammals, is negatively inhibited by the allosteric effector cytidine triphosphate, an end product of pyrimidine synthesis. The allosteric modulation occurs by the binding of effectors at the regulatory site of the enzyme. Noncovalent bonds are formed during the binding between effector and enzyme. In contrast, all the other enzymes are activated or deactivated by covalent modification. Chymotrypsinogen is secreted as an inactive proenzyme (zymogen) in pancreatic juice and is irreversibly activated by trypsin cleavage of a specific peptide bond. Glycogen phosphorylase is reversibly activated by phosphorylation of a specific serine residue. At the same time, glycogen synthase is reversibly deactivated by phosphorylation of a specific serine residue, thereby preventing a futile cycle of breakdown and resynthesis of glycogen. Pyruvate dehydrogenase also is reversibly inactivated by phosphorylation of a specific serine residue. In all four enzymes, a single, discrete, covalent modification leads to conformational changes that allow the switching on or off of enzyme activity.

61. The answer is A. *(Rawn, pp 53–60. Stryer, 4/e, pp 18–23.)* α-Amino acids contain an α-carboxyl group, an α-amino group, and a side chain or R group. All three of these groups are bound to the central carbon, or α-carbon, which is asymmetric. Amino acids are classified as acidic, neutral hydrophobic, neutral hydrophilic, or basic, depending on the charge or partial charge on the R group at pH 7.0. Basic R groups carry a positive charge at physiologic pH owing to protonated amide groups, while acidic R groups carry a negative charge owing to ionized carboxyl groups. Neutral hydrophilic side chains have uncharged but polar or partially charged groups. The structures shown in the question are (A) glycine, neutral; (B) γ-aminobutyric acid, neutral; (C) aspartic acid, acidic; (D) lysine, basic; and (E) threonine, neutral hydrophilic.

Glycine is the only α-amino acid shown with a hydrophobic or completely uncharged and nonpolar R group.

62. The answer is C. *(Stryer, 4/e, pp 367–368.)* The light chain and part of the heavy chain at the amino terminal contain the antibody-combining site in the "hypervariable regions." These regions are all contained in fragment A, which is known as *Fab.* Fragment B is known as *Fc.* The Fc fragment contains a site for binding of complement. The Fab fragments mediate complement fixation. Each of the fragments can be further dissociated into two subunits by breaking its disulfide bridge with mercaptoethanol or some other reducing agent.

63. The answer is E. *(Rawn, pp 133–135. Stryer, 4/e, pp 198–199.)* Unlike Michaelis-Menten enzymes, allosteric enzymes exhibit sigmoidal plots when reaction velocity is plotted against substrate concentrations. The enzyme contains both a catalytic site and a regulatory site. The binding of regulatory molecules to the regulatory site alters enzyme activity. The binding of one substrate molecule can affect the binding of substrate to other catalytic sites.

64. The answer is C. *(Stryer, 4/e, pp 24–25.)* To form proteins, the α-carboxyl groups of amino acids are joined to the α-amino groups of adjacent amino acids by a peptide (amide) bond. The formation of the peptide bond with a loss of a water molecule is shown below:

$$^{+}H_3N{-}\underset{R}{\overset{H}{C}}{-}\overset{O}{\overset{\|}{C}}{-}O^{-} + {}^{+}H_3N{-}\underset{R}{\overset{H}{C}}{-}\overset{O}{\overset{\|}{C}}{-}O^{-} \xrightarrow{\nearrow H_2O} {}^{+}H_3N{-}\underset{R}{\overset{H}{C}}{-}\overset{O}{\overset{\|}{C}}{-}\underset{H}{N}{-}\underset{R}{\overset{H}{C}}{-}\overset{O}{\overset{\|}{C}}{-}O^{-}$$

The peptide bond so formed has a partial double-bond character, being rigid and planar. Consequently, there is no free rotation around the bond between the carbonyl carbon and the amino nitrogen. Because of the lack of rotation around the bond, the peptide bond is usually a *trans* rather than a *cis* bond with the carbonyl oxygen being opposite the amino hydrogen. This is not the most thermodynamically stable configuration. The oxygen of the carbonyl group and the hydrogen of the amino group are partially charged and hence polar. Because of the partially charged nature of these groups, they are involved in forming hydrogen bonds such as the intrachain hydrogen bonds of the α-helices and β-sheets.

65. The answer is E. *(Stryer, 4/e, pp 647–650, 774.)* In humans, tyrosine can be formed by the hydroxylation of phenylalanine. This reaction is catalyzed by the enzyme phenylalanine hydroxylase. A deficiency of phenylalanine hydroxylase results in the disease called *phenylketonuria (PKU)*. In this disease it is usually the accumulation of phenylalanine and its metabolites rather than the lack of tyrosine that is the cause of the severe mental retardation ultimately seen. In contrast, it does not seem phenylalanine is formed by the removal of tyrosine's hydroxyl group to any great extent. Once formed, tyrosine is the precursor of many important signal molecules. Catalyzed by tyrosine hydroxylase, tyrosine is hydroxylated to form dopa, which in turn is decarboxylated to form dopamine in the presence of dopa decarboxylase. Then, norepinephrine and finally epinephrine are formed from dopamine. All of these are signal molecules to some degree. Dopa and inhibitors of the dopa decarboxylase are used in the treatment of Parkinson's disease, a neurologic disorder. Norepinephrine is a transmitter at smooth muscle junctions innervated by sympathetic nerve fibers. Epinephrine and dopamine are catecholamine transmitters synthesized in sympathetic nerve terminals and in the adrenal gland. Tyrosine is also the precursor of thyroxine, the major thyroid hormone, and melanin, a skin pigment.

66. The answer is E. *(Stryer, 4/e, pp 569–576, 593–595.)* Anabolic activities are those that occur in the well-fed state and are associated with high insulin levels. In the well-fed state, amino acid and glucose levels are high in the blood, stimulating increased secretion of insulin from the pancreas. Insulin enhances the removal of amino acids and glucose from the blood, thereby leading to their storage. Glucose is stored as glycogen in all tissues, but particularly in the liver, while amino acids are converted primarily to muscle proteins. Under these conditions, changes in blood glucose can be maintained by liver glycogen modulation, and gluconeogenesis is not needed. Gluconeogenesis only occurs in a time of need when glycogen stores are depleted. It is considered a catabolic reaction, even though it leads to the synthesis of glucose from glucogenic amino acids.

67. The answer is B. *(Stryer, 4/e, pp 18–23, 721–723.)* Except for threonine, which has an aliphatic side chain with a hydroxyl group, all the amino acids listed are sulfur-containing. Of these, only homocysteine is a sulfur-containing amino acid that is not found in proteins. It is an intermediate in the formation of cysteine from the sulfur-containing amino acid methionine. Homocysteine and methionine are also components of the activated methyl cycle in which *S*-adenosylmethionine is regenerated. *S*-Adenosylmethionine is one of the major donors of methyl groups. Methionine is an essential amino acid and must be derived from the diet.

68. The answer is A. *(Stryer, 4/e, pp 28–37.)* Regular arrangements of groups of amino acids located near each other in the linear sequence of a polypeptide are the secondary structure of a protein. The α-helix, β-sheet, and β-bend are the secondary structures usually observed in proteins. In both the α-helix and the β-sheet, all the peptide bond components participate in hydrogen bonding. That is, the oxygen components of the peptide bond form hydrogen bonds with the amide hydrogens. In the case of the α-helix, all hydrogen bonding is intrachain and stabilizes the helix. In the case of β-sheets, the bonds are interchain when formed between the polypeptide backbones of separate polypeptide chains and intrachain when the β-sheet is formed by a single polypeptide chain folding back on itself. While the spiral of the α-helix prevents the chain from being fully extended, the chains of β-sheets are almost fully extended and relatively flat. The chains of β-sheets can be either parallel or antiparallel. When the N terminals of chains run together, the chain or segment is considered parallel. In contrast, when N-terminal and C-terminal ends of the chains alternate, the β-strand is considered antiparallel.

69. The answer is C. *(Stryer, 4/e, pp 18–23.)* Glycine, the simplest of amino acids, contains only a single hydrogen as a side chain. The hydrogen is polar. The other amino acids listed have nonpolar side chains. Alanine has a single methylene group as its R group. Valine, leucine, and isoleucine are all branch chain amino acids. Valine has three methylene groups, while leucine and isoleucine have four.

70. The answer is C. *(Rawn, pp 40–44, 58–60. Stryer, 4/e, pp 23, 42–43.)* The figure in the question shows the titration curve of glycine, an amino acid with two dissociable protons—one from the α-carboxyl group and the other from the α-amino group. The maximum buffering capacity of any ionizable function is at the pH equivalent to the pK_a of the dissociation, as represented by points A and B on the graph.

71. The answer is A. *(Stryer, 4/e, pp 18–23.)* Except for terminal amino acids, all α-amino groups and all α-carboxyl groups are utilized in peptide bonds. Thus, only amino acids with side chains may be considered. Of these, 7 of the 20 common amino acids have easily ionizable side chains. These are the basic amino acids lysine, arginine, and histidine; the acidic amino acids aspartate and glutamate; as well as tyrosine and cysteine. Leucine is a branched amino acid with a hydrocarbon side chain of four carbons.

72. The answer is B. *(Stryer, 4/e, pp 35–39.)* The primary structure of a protein is simply its amino acid sequence and the location of disulfide bonds, if any are present. Thus, only procedures relevant to the determination of amino acid sequence are important to achieve this goal. Separation of original polypeptide chains as well as smaller fragments produced by experimental cleavage is carried out. The amino acid sequences of the small fragments are determined after fragmentation by different procedures. By comparing amino acid sequences of different sets of peptide fragments, with special attention given to areas of overlap, the original sequence can be determined. α-Helices and β-pleated sheets are examples of secondary structure of proteins.

73. The answer is E. *(Stryer, 4/e, pp 638–641.)* Following removal of the α-amino group from amino acids, the carbon skeletons are metabolized. Essentially all the carbon skeletons of amino acids are converted into major metabolic intermediates that can either be utilized to form glucose or oxidized in the citric acid cycle. Seven molecules are considered major intermediates into which amino acid carbon skeletons are converted: succinyl CoA, fumarate, oxaloacetate, α-ketoglutarate, pyruvate, acetyl CoA, and acetoacetyl CoA.

74. The answer is C. *(Stryer, 4/e, pp 18–23.)* The tetrahedral array of four different groups around the α-carbon results in optical activity in all amino acids except glycine. Two mirror forms of each amino acid are thus possible: L-isomers and D-isomers. Only L-isomers are utilized in proteins. Since glycine has two identical groups (hydrogen atoms) on the α-carbon, it is not optically active.

75. The answer is D. *(Stryer, 4/e, pp 372–376.)* Five classes of immunoglobulins are known: IgG, IgA, IgM, IgD, and IgE. The difference between each class is due to the variations in the constant chains from one class to another. The respective heavy chains corresponding to each class of immunoglobulin are γ—IgG; α—IgA; μ—IgM; δ—IgD; and ε—IgE. In contrast, the light chains are the same in each class: either κ or λ. The different biologic characteristics are due to the unique heavy chains. IgM is the first class of antibodies to be observed in the plasma following antigenic stimulation. IgG is the major antibody produced in serum at 10 days following antigenic stimulation. IgA acts against bacteria and viruses and is observed in external secretions such as mucus, tears, and saliva. While the undesirable effects of IgE in allergic reactions are known, its possible benefits are not understood. Likewise, the role of IgD is not known.

76. The answer is A. *(Stryer, 4/e, pp 247–251.)* Pepsinogen, procarboxypeptidase, chymotrypsinogen, and trypsinogen are inactive zymogen precursors of the gastrointestinal enzymes pepsin, carboxypeptidase, chymotrypsin, and trypsin, respectively. Ribonuclease is secreted in its active form. It catalyzes the depolymerization of ribonucleic acid.

77. The answer is D. *(Stryer, 4/e, pp 717–718.)* Of the 20 common amino acids, 9 are considered essential. These are histidine, isoleucine, leucine, lysine, methionine, phenylalanine, threonine, tryptophan, and valine. All other amino acids can be made from precursors by humans.

78. The answer is E. *(Rawn, pp 53–55. Stryer, 4/e, pp 18–23.)* Five of the twenty common amino acids have cyclic structures as side groups. The three-carbon side chain of proline is bonded to both the α-carbon and the amino group. Thus, proline is an amino acid. Histidine contains an imidazolium ring function with a positive charge. The three common aromatic acids are phenylalanine, tyrosine, and tryptophan. Tryptophan has an indole ring, tyrosine has a phenol ring, and phenylalanine has a benzyl ring. Each of the aromatic amino acids absorbs light significantly in the ultraviolet. Aspartate is an acidic amino acid with no cyclic function.

79. The answer is C. *(Stryer, 4/e, pp 247–255.)* Pepsinogen is the zymogen precursor of pepsin, one of the digestive proteases found in the stomach. Platelets are the cellular elements first activated at the site of disruption in blood vessels. Platelets agglutinate to form a plug and thereby temporarily prevent hemorrhage. Fibrinogen is the substrate acted upon by thrombin to yield the fibrin mesh of blood clots. Heparin is a mucopolysaccharide that terminates clot formation by interfering with a number of steps in the coagulation cascade. Once the structural integrity of the damaged area is repaired, plasmin hydrolyzes fibrin clots to form soluble products.

80. The answer is B. *(Rawn, pp 88–90. Stryer, 4/e, pp 31–32.)* The sequence and composition of collagen fibers are unique. Nearly every third amino acid residue is glycine. Proline, hydroxyproline, or proline and hydroxyproline in combination are often sequenced between repeating glycine residues. Proline and hydroxyproline compose about 20 percent of collagen. It is thought that the frequent proline residues determine the helical character of the collagen chain, while the glycine residues allow the chain to intertwine. Proline residues as well as a smaller proportion of lysine residues on the amino side of glycine residues are hydroxylated by prolyl- or lysyl-hydroxylase in the presence of the reducing agent ascorbic acid (vitamin C). Hydroxyproline and hydroxylysine residues are found in few other proteins. Isoleucine is common to many proteins.

81. The answer is E. *(Stryer, 4/e, pp 174–175.)* Many different mutations produce thalassemias. These genetic disorders are characterized by the defective synthesis of one or both of the hemoglobin chains. This may be contrasted to genetic diseases in which single amino-acid mutations produce normal quantities of hemoglobin chains with impaired functions, such as in sickle cell anemia. The absence or functional deficiency of hemoglobin can be brought about by these genetic disorders of hemoglobin synthesis. Deficiencies ranging from a missing gene to abnormal translation are responsible.

82. The answer is E. *(Stryer, 4/e, pp 198–199, 237–247, 493–495.)* Examples of allosteric enzymes include aspartate transcarbamoylase, which is inhibited by cytidine triphosphate (CTP); phosphofructokinase, which is inhibited by adenosine triphosphate (ATP) and activated by fructose 2,6-bisphosphate; hexokinase, which is inhibited by glucose-6-phosphate; and pyruvate kinase, which is inhibited by ATP. Allosteric enzymes produce sigmoidal kinetics when substrate concentration is plotted against reaction velocity. In contrast, hyperbolic plots are observed with Michaelis-Menten enzymes. The binding of effector molecules, such as end products or second messengers, to regulatory subunits of allosteric enzymes can either positively or negatively regulate catalytic subunits. Nonregulatory enzymes, such as lactate dehydrogenase, do not exhibit sigmoidal kinetics.

83. The answer is C. *(Rawn, pp 167–184. Stryer, 4/e, pp 192–195.)* The Michaelis-Menten equation is

$$v = \frac{V_{max}\,[S]_0}{K_m + [S]_0}$$

based on examination of the initial velocity (v) of an enzyme reaction where it can be assumed that the substrate concentration $[S]$ remained at the initial substrate concentration $[S]_0$ and that accumulation of product (P) was essentially zero. At high enough S, all the enzyme (E) is present as ES complex, and maximal velocity (V_{max}) is reached. At one-half V_{max}, the Michaelis constant, K_m (which is $[k_2 + k_3]/k_1$), represents the substrate concentration.

$$E + S \underset{k_2}{\overset{k_1}{\rightleftharpoons}} ES \overset{k_3}{\rightarrow} P + E$$

Since k_1, k_2, and k_3 usually are not measured, K_m must be determined empirically by measuring V_{max} and v at various substrate concentrations. Competitive inhibitors require a greater concentration of substrate to achieve a given velocity. Thus, K_m increases with competitive inhibition.

84. The answer is A. *(Stryer, 4/e, pp 474–475.)* Glycosaminoglycans are the polysaccharide chains of proteoglycans. These chains are made up of repeating units of disaccharides. The amino sugars forming the disaccharides contain negatively charged sulfate or carboxylate groups. The primary glycosaminoglycans found in mammals are hyaluronic acid, heparin, heparan sulfate, chondroitin sulfate, and keratan sulfate. Collagen is the major protein of connective tissue.

85. The answer is C. *(Rawn, pp 61–65. Stryer, 4/e, pp 45–50.)* At the isoelectric point of proteins, positive and negative charges balance. Proteins are least soluble at this point. In gel filtration columns, chromatographic separation of proteins is dependent upon the fact that large molecules move more rapidly through the column since only the smaller aqueous volume between beads is available due to their large size. Smaller molecules can distribute between the water phase in and out of the porous beads composing columns. Ammonium sulfate is so soluble in water that very high ionic strengths may be reached. As the concentration of ammonium sulfate is increased, different proteins in a mixture can be salted-out. The precipitated proteins produced by this procedure often retain biologic activity since they are not denatured by the process. In contrast, strong acids, like HCl, denature proteins as well as cause precipitation when used in a concentrated form (e.g., 4 *N*). Dilute acids, such as acetic acid, may actually solubilize certain proteins. Trichloroacetic acid is particularly good at deproteinizing solutions by precipitation. It binds to positively charged groups.

86. The answer is C. *(Stryer, 4/e, pp 190–191.)* Formation of an enzyme-substrate complex by the binding of the substrate to the active site of the enzyme is the first step in catalysis. This very specific reaction depends upon the precise arrangement of atoms in the active site, so that the substrate will be recognized and "fit." To this end, active sites are often very small, three-dimensional entities of the enzyme that usually involve amino acid residues from different regions of the linear amino acid sequence, not simply adjacent amino acids. These sites are usually clefts or crevices formed by the specific folding of the enzyme. The initial binding of substrate to the active site is a weak, noncovalent process.

87. The answer is C. *(Rawn, pp 53–54, 58–59. Stryer, 4/e, pp 18–23.)* In blood and other solutions at physiologic pH (approximately 7.0), only terminal carboxyl groups, terminal amino groups, and ionizable side chains of amino acid residues in proteins have charges. The basic amino acids lysine, arginine, and histidine have positive charges (protonated amines). The acidic amino acids aspartate and glutamate have negative charges (ionized carboxyls). Glutamine possesses an uncharged, but hydrophilic, side chain.

88. The answer is A. *(Rawn, pp 141–144. Stryer, 4/e, pp 168–173.)* The carboxyl of glutamate at position 6 on the β-chain of normal hemoglobin is dissociated and negatively charged at pH 7.0. Substitution of uncharged valine for glutamate by mutation produces sickle-cell hemoglobin, which is less negatively charged and has an increased electrophoretic mobility. Polymerization of the deoxygenated form of sickle hemoglobin occurs owing to the alteration of primary structure caused by the valine substitution. The insoluble, polymerized hemoglobin causes the erythrocyte to lose flexibility and become rigid and sickle-shaped. The brittle cells produce anemia and block capillaries.

89. The answer is D. *(Rawn, pp 101–111. Stryer, 4/e, pp 25–39.)* The primary structure usually determines the higher order structure of proteins, although there are certain well-known exceptions, such as insulin. Disulfide bonds are usually of more importance in stabilizing a protein's conformation once it has been attained than they are in determining it. Proteins are usually folded so that charged amino acids tend to be exposed to water, while hydrophobic side chains tend to be buried within the center of the molecule away from water. Since the quaternary structure of proteins refers to the assembly of two or more monomeric polypeptide chains into oligomeric proteins, not all proteins have a quaternary structure. Most secreted proteins are composed of only one polypeptide chain and have no quaternary structure.

90. The answer is B. *(Stryer, 4/e, pp 363–368.)* γ-Immunoglobulins are composed of two heavy and two light chains with intra- and interchain disulfide linkages. There are two antigen-binding sites per antibody molecule. The N terminal of both heavy and light chains is the variable region, whereas the C terminal is constant. Each potential antibody-making cell of newborns produces a single, specific antibody. This specificity is made before antigen is contacted and is the basis of the clonal selection process.

91. The answer is C. *(Rawn, pp 476, 485. Stryer, 4/e, p 723.)* In the synthesis of cysteine, the following sequence of steps occurs:

Adenosine
↑
Methionine → *S*-Adenosylmethionine → Homocysteine
B_6 ↓ ← Serine
Cysteine + α-Ketobutyrate + NH_4^+ ← Cystathionine

Cystathionine synthetase, a pyridoxal phosphate (B_6) enzyme, catalyzes the condensation of serine and homocysteine to form cystathionine. A deficiency of this enzyme leads to a buildup of homocysteine, which oxidizes to form homocystine. This usually results in mental retardation. The resulting cysteine deficiency must be made up from dietary sources. In some cases, dietary intake of vitamin B_6 (pyrixodal phosphate) may alleviate symptoms because of its requirement by the crucial enzymes. Likewise, inability to normally regenerate methionine from homocysteine by methylation of *N*-methyltetrahydrofolate will also cause a buildup of excess homocysteine. Homoserine levels would be high only under the normal conditions that favor cysteine synthesis.

92. The answer is A. *(Rawn, pp 135–141. Stryer, 4/e, pp 157–168.)* In addition to its function as a carrier of O_2 and CO_2, hemoglobin buffers sudden additions of acid or base to the blood by virtue of the histidine 146 on each β-chain. However, protonation of the imidazole of histidine causes deoxygenation of hemoglobin. Thus, decreased binding of O_2 occurs in the high pH conditions of acidosis. 2,3-Diphosphoglycerate (DPG) binds specifically to deoxyhemoglobin; that is, DPG cross-links positively charged residues on the β-chain, thereby decreasing oxygen affinity and stabilizing the deoxygenated form of hemoglobin. The addition of each O_2 molecule to deoxyhemoglobin requires the breakage of salt links, such as those formed by 2,3-DPG. Each subsequent O_2 molecule requires the breakage of fewer salt links. Thus, initial O_2 binding actually results in an increased affinity for subsequent O_2 binding, which in turn results in a cooperative allosteric binding mechanism. CO_2 reacts reversibly with the amino acid terminals of hemoglobin to create carbaminohemoglobin, which is negatively charged and which forms salt bridges stabilizing deoxyhemoglobin. Hence, CO_2 binding lowers the affinity of hemoglobin for O_2.

93. The answer is E. *(Rawn, pp 463–468. Stryer, 4/e, pp 634–636.)* Following the condensation of carbon dioxide and ammonia to form carbamoyl phosphate, the major compounds successively encountered in the urea cycle are ornithine, citrulline, aspartate, argininosuccinate, fumarate, arginine, and urea. Of these major compounds, only fumarate and urea do not meet the definition of an amino acid. An amino acid is defined as consisting of an α-carbon to which is attached an amino group, a carboxyl group, a hydrogen, and a distinct side group.

94–95. The answers are 94-E, 95-B. *(Stryer, 4/e, pp 35–39.)* The structure of a functional protein is defined by increasing levels of complexity. The primary structure of proteins is determined by the sequence of amino acids linked by peptide bonds. This sequence predicts secondary and tertiary structure. Secondary structure is the regular arrangement of amino acids close to one another in the sequence due to intrachain hydrogen bonds. This includes α-helix, β-sheet, and β-bend. Tertiary structure is the final folding of the protein into domains that arrange in a specific relationship to one another. The tertiary folding is caused and ultimately stabilized by bonds that include disulfide bonds, hydrogen bonds, electrostatic (ionic) interactions, and hydrophobic interactions. Chaperone, or helper, proteins may assist the folding of some proteins into the proper functional configuration. Quaternary structure only exists in functional proteins made up of two or more polypeptides held together by noncovalent interactions.

96–97. The answers are 96-D, 97-E. *(Stryer, 4/e, pp 648–650.)* Under normal conditions, tyrosine is synthesized from the essential amino acid phenylalanine by phenylalanine hydroxylase. In phenylketonuria (PKU), phenylalanine hydroxylase is absent. Phenylalanine is converted to phenylpyruvate and other aromatic compounds that accumulate in the blood and spill into the urine. The phenylketones damage the developing brain and produce mental retardation unless artificial protein lacking phenylalanine is fed to children with PKU. In this disease, tyrosine becomes an essential amino acid.

Under normal conditions, the essential branched-chain amino acids (isoleucine, leucine, and valine) are degraded by transamination to branched-chain, α-keto acids, which are oxidized and decarboxylated to form acyl CoA. Branched-chain, keto acid dehydrogenase is absent in maple syrup urine disease. Hence, the three α-keto acids derived from the branched-chain amino acids spill into the urine and give it a sweet smell. Artificial proteins low in branched-chain amino acids are fed to infants with the otherwise fatal disease.

98–99. The answers are 98-A, 99-B. *(Stryer, 4/e, pp 192–198.)* Substrate concentrations are usually expressed in terms of molarity, e.g., M = moles/liter, mM = millimoles/liter, μM = micromoles/liter. K_m, the Michaelis constant, is expressed in terms of substrate concentration. Each unit of enzyme activity is described as the amount of enzyme that converts a specific amount of substrate to a product within a given time. The standard units are micromoles per minute. Specific activity makes use of the units of enzyme activity that is being described as units per amount of protein or enzyme present.

100–103. The answers are 100-D, 101-E, 102-G, 103-B. *(Stryer 4/e, pp 18– 23.)* Each of the 20 unique amino acids coded for by DNA is composed of an α-carbon atom bonded to a hydrogen, a carboxyl group, an amino group, and a side-chain R group. The α-carbon is so named because it is adjacent to the acetic carboxyl group. The distinctive side chains of each different amino acid allow variation in charge, shape, size, and reactivity. Although glutamine is often referred to as an acidic amino acid, in fact, it is an uncharged polar amino acid with no ionizable group. It is an amide derivative of glutamate, which is an acidic amino acid with an ionizable carboxyl group. The cyclic amino (imino) acid proline is converted to hydroxyproline following incorporation into collagen by the enzyme prolyl hydroxylase. A small proportion of the lysine residues in collagen also are hydroxylated following incorporation. The reducing agent ascorbic acid (vitamin C) is necessary for this hydroxylation reaction. Hydroxylation occurs at specific sites on the collagen chains before they become helical. Aliphatic amino acids with large side chains, such as leucine, isoleucine, and valine, are hydrophobic in nature. Their hydrophobicity forces them to sequester together away from water in the interior of proteins. The three-dimensional structure of proteins is highly dependent on the hydrophobic side chains of aliphatic amino acids forming the interior of proteins. In contrast to aliphatic amino acids that have no ionizable side chains, basic amino acids have ionizable amino groups that are positively charged at neutral pH. These include lysine and arginine, which have a p*K* of pH 10 and pH 12, respectively, and histidine with an ionizable imidazole ring with a p*K* of 6.5. The different characteristics of the side chains of amino acids are responsible for the different qualities of the proteins into which they are incorporated.

104–107. The answers are 104-A, 105-B, 106-C, 107-G. *(Stryer, 4/e, pp 735–736.)* Reticuloendothelial cells degrade red blood cells following approximately 120 days in the circulation. The steps in the degradation of heme are as follows:

1. Formation of the green pigment biliverdin by the cleavage of the porphyrin ring of heme.

2. Formation of the red-orange pigment bilirubin by the reduction of biliverdin. The change in color of a bruise from bluish-green to reddish-orange reflects the heme degradation and the change in color of the bile pigments biliverdin and bilirubin.

3. Uptake of bilirubin by the liver and the formation of bilirubin diglucuronide. Bilirubin, which is quite insoluble, is transported to the liver attached to albumin. In the liver, bilirubin is conjugated to two glucuronic acid molecules to form bilirubin diglucuronide

4. Active excretion of bilirubin into bile. Once formed in the liver, bilirubin diglucuronide is transported against a concentration gradient into the bile. If bilirubin is not conjugated, it is not excreted.

5. Production of the color of stools and urine. Once the bile is excreted into the gut, bilirubin diglucuronide is hydrolyzed and reduced by bacteria to form urobilinogen, which is colorless. Much of the urobilinogen of the stools is further oxidized by intestinal bacteria to stercobilin, which gives stools their characteristic brown color. Some urobilinogen is reabsorbed by the gut into the portal blood, transported to the kidney, and converted and excreted as urobilin, which gives urine its characteristic yellow color.

6. Jaundice. Either a tremendous increase in the production of bilirubin, such as that found in hemolytic anemia or malaria; a blockage in the excretion of bilirubin due to liver damage; or an obstruction of the bile duct can lead to jaundice. Jaundice refers to the yellow color of the skin and eyes caused by the increased level of bilirubin in the blood.

108–109. The answers are 108-E, 109-A. *(Stryer, 4/e, pp 30–32, 474–478.)* Collagens are insoluble proteins that have great tensile strength. They are the main fibers composing the connective tissue elements of skin, bone, teeth, tendon, and cartilage. Collagen is composed of tropocollagen, a triple-stranded helical rod rich in glycine, proline, and hydroxyproline residues.

The second major macromolecular component of connective tissue is proteoglycan, which forms the so-called ground substance. Proteoglycans are composed of polysaccharide and protein. The polysaccharide chains are made up of repeats of negatively charged disaccharide units. This polyanionic quality of proteoglycans allows them to bind water and cations and, thus, determines the viscoelastic properties of connective tissues.

The cornified layer of epidermis derives its toughness and waterproof nature from keratin. Keratins are disulfide-rich proteins that compose the cytoskeletal elements known as intermediate filaments. Hair and animal horns are also composed of keratin.

110–111. The answers are 110-C, 111-A. *(Stryer, 4/e, pp 228, 247–252.)* Papain, the enzyme derived from papaya, has an active-site cysteine residue. It is in the thiol class of proteases. Such thiol enzymes form a covalent thioester bond between the acyl group of the substrate and the active-site sulfhydryl of the enzyme.

Carboxypeptidase A contains zinc at its active center. This is important during proteolysis since it causes electronic strain of the substrate, thereby inducing rearrangements of the electronic distribution that facilitate hydrolysis of carboxyl-terminal peptide bonds.

Most of the proteolytic enzymes from the pancreas are serine proteases. These include chymotrypsin, trypsin, and elastase. Thrombin and subtilisin also are serine proteases. Each of these enzymes has an unusually reactive serine residue at its active site that specifically reacts with diisopropyl phosphofluoridate and becomes inactivated.

Pepsin is an acid protease (carboxyl protease) that is only functional in the acid environment of the stomach. Most lysosomal proteases are acid proteases. Pancreozymin is not an enzyme. It is a polypeptide hormone, secreted by the gut, that stimulates pancreatic release of zymogens.

112–113. The answers are 112-A, 113-D. *(Stryer, 4/e, pp 192–195.)* Differential equations for the kinetics of enzyme-catalyzed reactions involve several parameters defined by the rate constants of such reactions. For the reaction $S + E \underset{k_2}{\overset{k_1}{\rightleftharpoons}} S \cdot E$, the dissociation constant K_s equals the Michaelis constant, inasmuch as k_3 is absent and the expression $\frac{k_2 + k_3}{k_1}$ becomes $\frac{k_2}{k_1}$. Where the enzyme-substrate complex proceeds irreversibly at rate k_3 to yield product P in the equation reproduced in the question, the reaction velocity, which is proportional to enzyme-substrate concentration (ES), is expressed as $V = \frac{k_3 ES}{K_m + S}$.

Maximal velocity (V_{max}) obtains at initial enzyme concentration (E_t), and for a constant substrate concentration is defined as $V = k_3 E_t$.

Nucleic Acids

DIRECTIONS: Each question below contains five suggested responses. Select the **one best** response to each question.

114. Restriction fragment length polymorphism (RFLP) analysis can only be used to follow the inheritance of a genetic disease if

(A) mRNA probes are used in combination with antibodies
(B) the disease-causing mutation is at or closely linked to an altered restriction site
(C) proteins of mutated and normal genes migrate differently upon gel electrophoresis
(D) mutations are outside of restriction sites so that cleaving still occurs
(E) restriction fragments remain the same size but their charge changes

115. Which of the following statements accurately describes sex hormones?

(A) They bind specific membrane receptors
(B) They interact with DNA directly
(C) They cause release of a proteinaceous second messenger from the cell membrane
(D) They enhance transcription when bound to receptors
(E) They inhibit translation through specific cytoplasmic proteins

116. If the graph shown below represents DNA containing a normal mixture of bases, it can be expected that the graph will be shifted to the left for duplex DNA containing a high content of

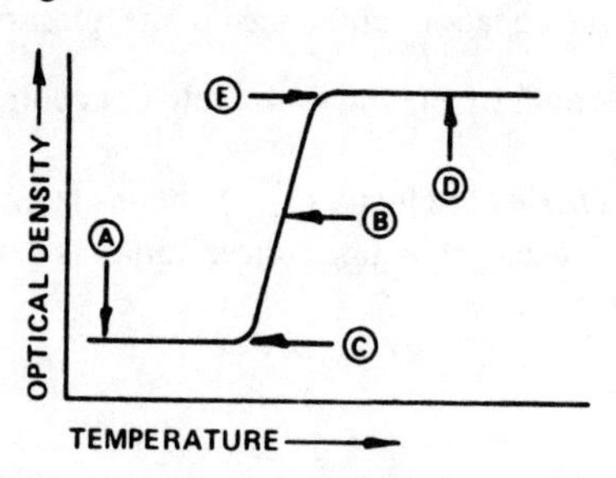

(A) adenine plus cytosine
(B) cytosine plus guanine
(C) adenine plus thymine
(D) cytosine plus thymine
(E) adenine plus guanine

117. An unknown enzyme that affects DNA has been isolated from *E. coli.* When a solution of this enzyme is mixed with supercoiled plasmid DNA, its only effect is to relax the DNA. At the end of the exposure to the enzyme solution, the plasmid DNA is covalently closed and still circular. This enzyme is

(A) restriction endonuclease
(B) primase
(C) reverse transcriptase
(D) helacase
(E) topoisomerase

118. It is well known that DNA polymerases synthesize DNA only in the 5′ to 3′ direction. Yet, at the replication fork, both strands of parental DNA are being replicated with the synthesis of new DNA. How is it possible that while one strand is being synthesized in the 5′ to 3′ direction, the other strand appears to be synthesized in the 3′ to 5′ direction? This apparent paradox is explained by

(A) 3′ to 5′ replicase
(B) 3′ to 5′ antipolymerase
(C) Okazaki fragments
(D) immediate replication and transfer of the leading strand
(E) lack of RNA primer on one of the strands

119. In contrast to DNA polymerase, RNA polymerase

(A) fills in the gap between Okazaki fragments
(B) works only in a 5′ to 3′ direction
(C) edits as it synthesizes
(D) synthesizes RNA primer to initiate DNA synthesis
(E) adds nucleoside monophosphates to the growing polynucleotides

120. What is the correct order of the following steps in protein synthesis?

1. A peptide bond is formed.
2. The small ribosomal subunit is loaded with initiation factors, messenger RNA, and initiation aminoacyl-transfer RNA.
3. The intact ribosome slides forward three bases to read a new codon.
4. The primed small ribosomal subunit binds with the large ribosomal subunit.
5. Elongation factors deliver aminoacyl-tRNA to bind to the A site.

(A) 1, 2, 3, 4, 5
(B) 2, 3, 4, 5, 1
(C) 4, 5, 1, 3, 2
(D) 3, 2, 4, 5, 1
(E) 2, 4, 5, 1, 3

121. Which of the following statements regarding repressor proteins is true?

(A) They inhibit transcription
(B) They bind to catabolite activator protein
(C) They bind to a regulatory gene
(D) They are produced by the operator site
(E) They act independently on DNA

122. Ultraviolet light causes DNA damage by

(A) activating photoreacting enzymes
(B) causing pyrimidine dimers
(C) activating DNA photolyase
(D) methylating purines
(E) cutting phosphodiester bonds

123. The removal of introns and subsequent self-splicing of adjacent exons occurs in some portions of primary ribosomal RNA transcripts. The splicing of introns in messenger RNA precursors is

(A) RNA-catalyzed in the absence of protein
(B) self-splicing
(C) carried out by spliceosomes
(D) controlled by RNA polymerase
(E) regulated by RNA helicase

124. Given that the chromosomes of mammalian cells may be 20 times as large as those of *E. coli,* how can replication of mammalian chromosomes be carried out in just a few minutes?

(A) Eukaryotic DNA polymerases are extraordinarily fast compared with prokaryotic polymerases
(B) The higher temperature of mammalian cells allows for an exponentially higher replication rate
(C) Hundreds of replication forks work simultaneously on each piece of chromosomal DNA
(D) A great many different RNA polymerases carry out replication simultaneously on chromosomal DNA
(E) The presence of histones speeds up the rate of chromosomal DNA replication

125. Which one of the following has catalytic activity?

(A) Nucleotides
(B) Phospholipids
(C) Glycogen
(D) RNA
(E) Collagen

126. An expression library of human DNA in *E. coli* can be screened by

(A) nitrocellulose blotting
(B) eastern blotting
(C) northern blotting
(D) Southern blotting
(E) western blotting

127. The Ames test for potential chemical carcinogens

(A) utilizes nude mice
(B) determines mutagenic potential by measuring reversion of bacterial mutations
(C) requires back transplantation in genetically related rats
(D) measures mutagenicity by computer modeling and prediction
(E) measures transformation in cultures of mammalian cells

128. Some of the enzymes utilized in DNA replication are (1) DNA-directed DNA polymerase, (2) unwinding proteins, (3) DNA polymerase I, (4) DNA-directed RNA polymerase, and (5) DNA ligase. What is the correct sequence of their use during DNA synthesis?

(A) 4, 3, 1, 2, 5
(B) 2, 3, 4, 1, 5
(C) 4, 2, 1, 5, 3
(D) 4, 2, 1, 3, 5
(E) 2, 4, 1, 3, 5

129. Which of the following statements regarding a double-helical molecule of DNA is true?

(A) All hydroxyl groups of pentoses are involved in linkages
(B) Bases are perpendicular to the axis
(C) Each strand is identical
(D) Each strand is parallel
(E) Each strand replicates itself

130. *S*-Adenosylmethionine is shown below with five substituent groups labeled A through E. Which group is *S*-adenosylmethionine able to donate in creatine synthesis?

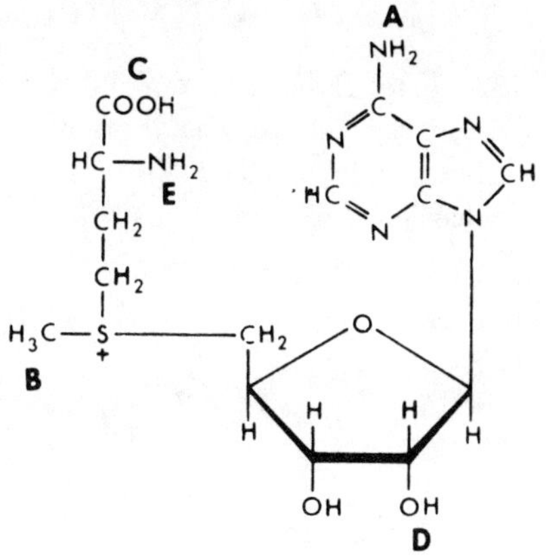

(A) A
(B) B
(C) C
(D) D
(E) E

131. Which of the following statements regarding eukaryotic cells is true?

(A) Formylated methionyl tRNA is important for initiation of translation
(B) Single mRNAs specify more than one gene product
(C) Cycloheximide blocks elongation during translation
(D) Cytosolic ribosomes are smaller than those found in prokaryotes
(E) Erythromycin inhibits elongation during translation

Questions 132–133

```
C5    0
|  /     \
C4        C1
  \      /
   C3—C2
```

132. In the carbon skeleton of the pentose sugar above, what is the attachment point of a base to form a nucleoside?

(A) C1
(B) C2
(C) C3
(D) C4
(E) C5

133. Which carbon of the pentose is in ester linkage with the phosphate in a nucleotide structure?

(A) C1
(B) C2
(C) C3
(D) C4
(E) C5

134. A promoter site on DNA

(A) transcribes repressor
(B) initiates transcription
(C) codes for RNA polymerase
(D) regulates termination
(E) translates specific proteins

135. The mushroom poison amanitin is an inhibitor of the synthesis of

(A) protein
(B) mRNA
(C) DNA
(D) glycoprotein
(E) adenosine

136. The following double-stranded polynucleotide is diagramed with its terminal ends shown and with a relatively long midpiece indicated as a dashed line. The terminal ends are single-stranded where the bases are indicated.

```
CCTAGG---------
      ----------GGATCC
```

Which of the following statements is true?

(A) Terminal repetition is not observed
(B) The end sequences are palindromes
(C) The polynucleotide is similar to intact plasmid DNA
(D) 5′-Exonuclease could not have produced the polynucleotide
(E) The polynucleotide segment can form a circular piece of DNA

137. An increased melting temperature (T_m) for duplex DNA results from a high content of

(A) adenine (A) + guanine (G)
(B) cytosine (C) + thymine (T)
(C) adenine + thymine
(D) cytosine + guanine
(E) adenine + cytosine

138. During DNA replication, the sequence 5′-TpApGpAp-3′would produce which of the following complementary structures?

(A) 5′-TpCpTpAp-3′
(B) 5′-ApTpCpTp-3′
(C) 5′-UpCpUpAp-3′
(D) 5′-GpCpGpAp-3′
(E) 3′-TpCpTpAp-3′

139. AUG, the only identified codon for methionine, is important as

(A) the site of attachment for the 30S ribosomal particle
(B) the recognition site on the transfer RNA
(C) a releasing factor for peptide chains
(D) a chain-terminating codon
(E) a chain-initiating codon

140. Modification of mRNA so that a signal sequence is added to the amino terminus of the cytosolic protein, α-globin, results in

(A) no change in physiology of the protein
(B) proteolytic cleavage within the cytosol
(C) translocation across the endoplasmic reticulum
(D) cytosolic localization of the protein
(E) signal recognition particle synthesis

141. Which of the following mutations is most likely to be lethal?

(A) Substitution of adenine for cytosine
(B) Substitution of cytosine for guanine
(C) Substitution of methylcytosine for cytosine
(D) Deletion of three nucleotides
(E) Insertion of one nucleotide

142. DNA replicates in a semiconservative manner. If a completely radioactive double-stranded DNA molecule undergoes two rounds of replication in a solution free of radioactive label, what is the radioactivity status of the resulting four DNA molecules?

(A) Half should contain no radioactivity
(B) All should contain radioactivity
(C) Half should contain radioactivity in both strands
(D) One should contain radioactivity in both strands
(E) None should contain radioactivity

143. How many high-energy phosphate-bond equivalents are utilized in the process of activation of amino acids for protein synthesis?

(A) Zero
(B) One
(C) Two
(D) Three
(E) Four

144. Sickle cell anemia is the clinical manifestation of homozygous genes for an abnormal hemoglobin molecule. The mutational event responsible for the mutation in the β-chain is

(A) crossing over
(B) insertion
(C) deletion
(D) nondisjunction
(E) point mutation

145. The hydrolytic step leading to the release of a polypeptide chain from a ribosome is catalyzed by

(A) stop codons
(B) peptidyl transferase
(C) release factors
(D) dissociation of ribosomes
(E) UAA

146. Thymine is present in which of the following?

(A) Ribosomal RNA
(B) Prokaryotic mRNA
(C) Mammalian mRNA
(D) tRNA
(E) None of the above

147. Sigma factor is best described as a

(A) subunit of RNA polymerase responsible for the specificity of the initiation of transcription of RNA from DNA
(B) subunit of DNA polymerase that allows for synthesis in both $5' \rightarrow 3'$ and $3' \rightarrow 5'$ directions
(C) subunit of the 50S ribosome that catalyzes peptide bond synthesis
(D) subunit of the 30S ribosome to which mRNA binds
(E) factor that forms the bridge between the 30S and 50S particles constituting the 70S ribosome

148. The function of signal recognition particles is to

(A) cleave signal sequences
(B) detect cytosolic proteins
(C) direct the signal sequences to ribosomes
(D) bind ribosomes to endoplasmic reticulum
(E) bind mRNA to ribosomes

149. In contrast to eukaryotic mRNA, prokaryotic mRNA

(A) can be polycistronic
(B) has a poly A tail
(C) is synthesized with introns
(D) requires splicing enzyme
(E) has 7-methylguanosine at the 5′ end

150. Tetracycline prevents synthesis of polypeptides by

(A) competing with mRNA for ribosomal binding sites
(B) blocking mRNA formation from DNA
(C) releasing peptides from mRNA-tRNA complexes
(D) inhibiting tRNA binding to mRNA
(E) preventing amino acid binding by tRNA

151. In prokaryotes, chloramphenicol has which of the following biochemical effects?

(A) It causes premature release of the polypeptide chain
(B) It causes misreading of the mRNA
(C) It depolymerizes DNA
(D) It inhibits protein synthetic activities of the 30S ribosomal subunit
(E) It inhibits protein synthetic activities of the 50S ribosomal subunit

152. Owing to "wobble," which of the following occurs?

(A) Strict Watson-Crick base-pairing of codons and anticodons occurs
(B) Certain anticodons pair with codons differing at the 3′ end
(C) Certain codons pair with anticodons differing at the 5′ end
(D) Lysogenic virus can become virulent
(E) Errors in transcription occur

153. Which of the following statements about ribosomes is true?

(A) They are an integral part of transcription
(B) They are found both free in the cytoplasm and bound to membranes
(C) They are bound together so tightly they cannot dissociate under physiologic conditions
(D) They are composed of RNA, DNA, and protein
(E) They are composed of three subunits of unequal size

154. A culture of bacteria not resistant to tetracycline develops an infection from a virus that is derived from the lysis of tetracycline-resistant bacteria. Most of the bacterial progeny of the original culture is found to have become resistant to tetracycline. What phenomenon has occurred?

(A) Conjugation
(B) Colinearity
(C) Recombination
(D) Transformation
(E) Transduction

155. Following ultraviolet damage of DNA in skin,

(A) a specific excinuclease detects damaged areas
(B) purine dimers are formed
(C) both strands are cleaved
(D) endonuclease removes the strand
(E) DNA hydrolysis does not occur

156. So-called caps of RNA molecules

(A) allow tRNA to be processed
(B) occur at the 3′end of tRNA
(C) are composed of poly A
(D) are unique to eukaryotic mRNA
(E) allow correct translation of prokaryotic mRNA

157. Transcription of the following sequence of the tryptophan operon occurs in the direction indicated by the arrow. What would be the base sequence of the mRNA produced?

3′. . .CGCCGCTGCGCG. . .5′

→ mRNA

5′. . .GCGGCGACGCGC. . .3′

(A) 5′. . .GCGGCGACGCGC. . .3′
(B) 5′. . .GCGCGUCGCCGC. . .3′
(C) 5′. . .GCGCGTGCGGCG. . .3′
(D) 5′. . .GCGGCGUCGCGC. . .3′
(E) 5′. . .CGCGCTCGCCGC. . .3′

---

158. Carbamoyl phosphate synthetase functioning in cytosol, but not mitochondria,

(A) is inhibited by uridine monophosphate
(B) is activated by acetylglutamate
(C) has a high activity
(D) cannot be found in liver
(E) is involved in purine biosynthesis

159. Guanosine triphosphate (GTP) is required by which of the following steps in protein synthesis?

(A) Aminoacyl-tRNA synthetase activation of amino acids
(B) Attachment of ribosomes to endoplasmic reticulum
(C) Translocation of tRNA–nascent protein complex from the A to P sites
(D) Attachment of mRNA to ribosomes
(E) Attachment of signal recognition protein to ribosomes

160. The function of factor rho is to

(A) bind catabolite repressor to the promoter region
(B) increase the rate of RNA synthesis
(C) eliminate the binding of RNA polymerase to the promoter
(D) participate in the proper termination of transcription
(E) allow proper initiation of transcription

161. Which of the following statements regarding the TATA box (Hogness box) is true?

(A) It is located on the first structural gene
(B) It binds RNA polymerase
(C) It encodes repressor protein
(D) It terminates protein synthesis
(E) It binds to anticodon

162. The steps in bacterial translation include (1) association of mRNA, initiation factors, and ribosomal subunits; (2) activation of amino acids; (3) formation of a peptide bond; (4) translocation of peptidyl-tRNA; and (5) binding together of GTP, elongation factor, and aminoacyl-tRNA. What is the correct chronologic sequence of these events?

(A) 1,5,2,3,4
(B) 2,1,5,3,4
(C) 5,1,2,4,3
(D) 2,1,5,4,3
(E) 1,5,2,4,3

163. Diphtheria toxin is often lethal in unimmunized persons because it

(A) inhibits initiation of protein synthesis by preventing the binding of GTP to the 40S ribosomal subunit
(B) binds to the signal recognition particle receptor on the cytoplasmic face of the endoplasmic reticulum receptor
(C) shuts off signal peptidase
(D) blocks elongation of proteins by inactivating elongation factor 2 (EF2, or translocase)
(E) causes deletions of amino acid by speeding up the movement of peptidyl-tRNA from the A site to the P site

164. Which of the following statements correctly describes eukaryotic nuclear chromosomal DNA?

(A) Each discontinuous piece making up the chromosomes of eukaryotes is about the same size as each prokaryotic chromosome
(B) Unlike bacterial DNA, no histones are associated with it
(C) It is not replicated semiconservatively
(D) It is a linear and unbranched molecule
(E) It is not associated with a specific membranous organelle

165. Xeroderma pigmentosum is an inherited human skin disease that causes a variety of phenotypic changes in skin cells exposed to sunlight. The molecular basis of the disease appears to be

(A) rapid water loss caused by defects in the cell membrane permeability
(B) the inactivation of temperature-sensitive transport enzymes in sunlight
(C) the induction of a virulent provirus on ultraviolet exposure
(D) the inability of the cells to synthesize carotenoid-type compounds
(E) a defect in an excision-repair system that removes thymine dimers from DNA

166. A potent inhibitor of protein synthesis that acts as an analogue of aminoacyl-tRNA is

(A) mitomycin C
(B) streptomycin
(C) nalidixic acid
(D) rifampicin
(E) puromycin

167. Which of the following statements correctly describes the nucleolus of a eukaryotic cell?

(A) It differs from that found in bacterial cells in that histones are present
(B) It may contain hundreds of copies of genes for different types of ribosomal RNAs
(C) It synthesizes 5S ribosomal RNA
(D) It synthesizes 60S and 40S ribosomal subunits
(E) It synthesizes all ribosomal RNA primary transcripts

168. The formation of thymine dimers as a result of sunburn

(A) requires an enzyme repair mechanism that utilizes ligase
(B) does not affect pyrimidines
(C) creates covalent bonds between thymines on opposite nucleotide strands
(D) is transcribed as frameshift mutations
(E) is repaired by thymidine hydroxylase

169. A signal sequence of amino acid residues is required for which of the following reactions?

(A) The synthesis of cytosolic proteins
(B) The synthesis of polyribosomal units
(C) The attachment of ribosomes to mRNA
(D) The attachment of ribosomes to endoplasmic reticulum
(E) The release of peptides from ribosomes

170. In studies of the mechanism of bacterial DNA replication, 5-bromouracil often is used as an analogue of thymidine in order to

(A) cause specific frameshift mutations for sequencing studies
(B) stop DNA synthesis at sites of thymidine incorporation
(C) provide a reactive group in the DNA affinity supports
(D) synthesize a denser DNA that can be identified by centrifugation
(E) create specific sites in the DNA for mild chemical cleavage

171. Which one of the following single-stranded DNA molecules would be palindromic in the double-stranded state?

(A) A-T-G-C-C-G-T-A
(B) A-T-G-C-T-A-C-G
(C) G-T-C-A-T-G-A-C
(D) G-T-A-T-C-T-A-T
(E) G-C-T-A-T-G-A-C

172. In the following partial sequence of mRNA, a mutation of the template DNA results in a change in codon 91 to UAA. What type of mutation is it?

88	89	90	91	92	93	94
GUC	GAC	CAG	UAG	GGC	UAA	CCG

(A) Missense
(B) Silent
(C) Nonsense
(D) Suppressor
(E) Frameshift

173. Specific radioactive identification of ribosomal RNA can be achieved by using ^{14}C-labeled

(A) cytosine
(B) guanine
(C) uracil
(D) deoxyribose
(E) thymine

174. Which one of the following statements correctly describes the synthesis of eukaryotic RNA?

(A) Encoding genes cannot be continuous
(B) Identical mRNA molecules may be derived from different copies of a specifically repeated gene
(C) Most genes produce primary mRNA that is continuous
(D) All genes coding for mRNA are continuous
(E) No genes coding for mRNA are discontinuous

175. *Kilobase* refers to

(A) a 1000-molecular-weight piece of double-stranded DNA
(B) 1000 base pairs of a double-stranded DNA molecule
(C) a 1000-molecular-weight piece of any DNA molecule
(D) 1000 base pairs of messenger RNA
(E) a 1000-molecular-weight piece of any RNA type

176. Complementary DNA is synthesized

(A) from retroviral RNA
(B) from messenger RNA
(C) from plasmid DNA
(D) using oligo dG primer
(E) using bacterial RNA

177. Studies of the genetic code in prokaryotes have revealed that

(A) messenger RNA molecules specify only one polypeptide chain
(B) many triplets can be "nonsense" triplets
(C) no signal exists to indicate the end of one codon and the beginning of another
(D) the nucleotide on the 5' end of a triplet has the least specificity for an amino acid
(E) gene sequence and encoded proteins are not colinear

178. Which one of the following statements correctly describes eukaryotic DNA?

(A) It uses DNA polymerase with nuclease activities
(B) It is replicated bidirectionally at many points
(C) It contains no repetitive DNA
(D) It is nonlinear
(E) It is found in a single helix form in chromosomes

179. Which one of the following causes a frameshift mutation?

(A) Transition
(B) Transversion
(C) Deletion
(D) Substitution of purine for pyrimidine
(E) Substitution of pyrimidine for purine

180. Which one of the following contributes nitrogen atoms to both purine and pyrimidine rings?

(A) Aspartate
(B) Carbamoyl phosphate
(C) Carbon dioxide
(D) Glutamine
(E) Tetrahydrofolate

181. Using written convention, which one of the following sequences is complementary to TGGCAGCCT?

(A) ACCGTCGGA
(B) ACCGUCGGA
(C) AGGCTGCCA
(D) TGGCTCGGA
(E) TCCGACGGT

182. Which one of the following binds to specific nucleotide sequences that are upstream and most distant from the start site?

(A) RNA polymerase
(B) Repressor
(C) Inducer
(D) Histone
(E) Restriction endonuclease

183. Ribosomes similar to those of bacteria are found in

(A) plant nuclei
(B) cardiac muscle cytoplasm
(C) pancreatic mitochondria
(D) liver endoplasmic reticulum
(E) neuronal cytoplasm

184. DNA is the major component of

(A) retrovirus
(B) repressor
(C) inducer
(D) operon
(E) histone

DIRECTIONS: Each numbered question or incomplete statement below is NEGATIVELY phrased. Select the **one best** lettered response.

185. All the following statements describe both the spiral structure of double-stranded DNA and the spiral structure found in certain segments of protein EXCEPT

(A) they are repeating spiral structures
(B) the units are specifically arranged in polymeric chains
(C) they have hydrogen bonding
(D) they are α-helical
(E) they have covalently linked backbones

186. The mechanisms of synthesis of DNA and RNA are similar in all the following ways EXCEPT

(A) they involve release of pyrophosphate from each nucleotide added
(B) they require activated nucleotide precursor and magnesium ion
(C) the direction of synthesis is $5' \rightarrow 3'$
(D) base pairing determines the correct sequence of bases added
(E) they require a primer

187. Template-directed DNA synthesis occurs in all the following EXCEPT

(A) the replication fork
(B) growth of RNA tumor viruses
(C) polymerase chain reactions
(D) expression of oncogenes
(E) repair of thymine dimers

188. All the following descriptions of DNA replication are common to the synthesis of both leading and lagging strands EXCEPT

(A) RNA primer is synthesized
(B) DNA polymerase III synthesizes DNA
(C) helicase (rep protein) continuously unwinds duplex DNA at the replication fork during synthesis
(D) nucleoside monophosphates are added in a 5′to 3′direction along the growing DNA chain
(E) DNA ligase repeatedly joins the ends of DNA along the growing strand

189. The following statements are true of all transfer (t) RNAs EXCEPT

(A) the 5′end is phosphorylated
(B) they are single chains
(C) methylated bases are found
(D) the anticodon loop is identical
(E) the 3′-end base sequence is CCA

190. All the following statements describing mammalian nuclear DNA during interphase are correct EXCEPT that

(A) the base ratio cytosine + thymine/guanine + adenine equals 1
(B) the number of sugar units equals the number of base units
(C) three hydrogen bonds are present between adenine-thymine pairs
(D) thymine and adenine contents are similar
(E) the number of pyrimidines equals the number of purines

191. All the following statements regarding the nucleic acid DNA are true EXCEPT

(A) it shows homology in closely related species
(B) it can be present as a single strand in some viruses
(C) it is found outside of chromosomes in bacteria
(D) it is the genetic material of all viruses
(E) it codes for RNA in mitochondria

192. All the following statements describing restriction endonucleases are true EXCEPT

(A) they do not yield single-stranded ends on complementary pieces of DNA
(B) they are restricted by methylation of recognition sequences
(C) they recognize palindrome sequences
(D) they cleave both strands in duplex DNA
(E) they are specific for short symmetric sequences

193. All the following are characteristic of the DNA-sequencing technique of Maxam and Gilbert EXCEPT

(A) specific cleavage of DNA at each base
(B) western blotting
(C) insertion of ^{32}P at a specific strand terminus
(D) autoradiography
(E) gel electrophoresis

194. All the following descriptions of eukaryotic RNA are true EXCEPT

(A) it derives from antiparallel, complementary DNA
(B) RNA is normally single-stranded
(C) units of uridine equal units of adenine
(D) it has an overall negative charge at neutral pH
(E) the ratio of ribose to base equals one

195. Peptide-chain elongation involves all the following EXCEPT

(A) peptidyl transferase
(B) GTP
(C) Tu, Ts, and G factors
(D) formylmet-tRNA
(E) mRNA

196. During protein synthesis, the formation of each peptide bond requires the equivalent of four high-energy phosphate bonds. Consumption of a high-energy phosphate bond is required in all the following steps EXCEPT

(A) peptidyl transferase formation of peptide bond
(B) binding of aminoacyl-tRNA to the active site of the ribosome
(C) translocation
(D) activation of amino acids
(E) binding of *N*-formylmethionyl-tRNA to the initiation codon of mRNA and association of ribosomal subunits

197. *N*-Formylmethionyl-tRNA is associated with synthesis of all the following EXCEPT

(A) chloroplast ribosomes
(B) mitochondrial ribosomes
(C) eukaryotic cytoplasmic ribosomes
(D) *E. coli* ribosomes
(E) *E. coli* proteins

198. All the following statements apply to the enzyme DNA polymerase III EXCEPT that

(A) deoxyribonucleotides are added to the 5′-hydroxyl terminus of preexisting DNA
(B) deoxyribonucleotides are added to a primer chain
(C) four different deoxyribonucleoside 5′-triphosphates are required
(D) double-stranded DNA may be used as a template
(E) pyrophosphate is a product of the reaction

199. All the following statements apply to the artificial cloning of DNA EXCEPT that

(A) specific restriction endonucleases may be used to split plasmid and donor DNA molecules at unique complementary sites
(B) vehicle DNA and donor DNA annealed together result in recombinant DNA
(C) in addition to plasmids, λ-bacteriophage may be used as a vector for forming recombinant DNA
(D) selection of bacteria containing recombinant DNA is often based on using vectors conferring antibiotic resistance
(E) the introduction of recombinant DNA molecules into host cells requires conjugation

200. All the following drugs inhibit the formation of deoxythymidylate EXCEPT

(A) methotrexate
(B) aminopterin
(C) fluorouracil
(D) chloramphenicol

201. True descriptions of direct effects of the AIDS drug azidothymidine (AZT) include all the following EXCEPT

(A) it prevents viral replication
(B) it inhibits RNA synthesis
(C) it inhibits DNA polymerase
(D) it inhibits DNA provirus production
(E) it inhibits reverse transcriptase

202. In the genetic code, UCC and UCU code for Ser; CUU codes for Leu; UUC codes for Phe; and CCU codes for Pro. Assume that synthesis may be initiated at any base along the mRNA. All the following dipeptides may be coded for by the mRNA sequence UCCUUCUUC EXCEPT

(A) Phe-Phe
(B) Leu-Leu
(C) Pro-Leu
(D) Pro-Ser
(E) Ser-Phe

203. DNA ligase is essential for all the following EXCEPT

(A) DNA repair
(B) formation of phosphodiester bonds between DNA chains of double-helical DNA molecules
(C) formation of phosphodiester bonds between DNA chains of single-stranded DNA molecules
(D) formation of a phosphodiester bond between the 5′-phosphate group at the end of one DNA chain and the 3′-hydroxyl group at the end of the other chain
(E) normal DNA synthesis

204. All the following are true statements about what xanthine can be EXCEPT

(A) a direct precursor of guanine
(B) a substrate of the enzyme xanthine oxidase
(C) a product of the enzyme xanthine oxidase
(D) oxidized to form uric acid
(E) a result of the oxidation of hypoxanthine

205. Feedback inhibition of pyrimidine nucleotide synthesis can occur by all the following means EXCEPT

(A) regulation of carbamoyl phosphate synthetase
(B) regulation of aspartate transcarbamoylase
(C) CTP allosteric effects
(D) UMP allosteric effects
(E) TTP allosteric effects

206. All the following are true statements about translation EXCEPT

(A) the genetic code can be overlapping
(B) the last nucleotide in a codon has less specificity than the others
(C) more than one group of nucleotides may code for a single amino acid
(D) three nucleotide bases code for one amino acid
(E) specific nucleotide sequences signal termination of peptide chains

207. Purine nucleotide biosynthesis can be inhibited by all the following EXCEPT

(A) GMP
(B) UMP
(C) AMP
(D) IMP
(E) blockage of IMP metabolism

208. Protein biosynthesis requires all the following EXCEPT

(A) ribosomal RNA
(B) messenger RNA
(C) primer protein
(D) peptidyl transferase
(E) GTP

209. The sex-linked, recessive absence of hypoxanthine-guanine phosphoribosyltransferase (HGPRT) in Lesch-Nyhan syndrome may cause all the following EXCEPT

(A) gout
(B) mental retardation
(C) decreased hypoxanthine levels
(D) increased serum urate levels
(E) increased phosphoribosylpyrophosphate levels

210. All the following statements correctly describe the recombinant DNA tool known as *plasmids* EXCEPT

(A) they occur normally in certain strains of bacteria
(B) they are circular DNA duplexes
(C) they are a class of mobile genetic elements (transposons)
(D) they do not affect drug resistance
(E) they do not prevent genetic rearrangement of nonhomologous loci of DNA molecules

211. With a single nucleotide substitution in DNA coding for the primary structure of a protein, transcription and translation could ultimately yield all the following EXCEPT

(A) a normally functioning protein with one amino acid change
(B) a protein with no function and one amino acid change
(C) a protein with altered function and one amino acid change
(D) a protein with normal function and normal amino acid composition
(E) no measurable protein production

212. All the following compose nucleosides EXCEPT

(A) phosphate groups
(B) deoxyribose sugar
(C) ribose sugar
(D) pentose sugar
(E) purine base

DIRECTIONS: Each group of questions below consists of lettered headings followed by a set of numbered items. For each numbered item select the **one** lettered heading with which it is **most** closely associated. Each lettered heading may be used **once, more than once, or not at all.**

Questions 213–214

Match each base with the formula describing it.

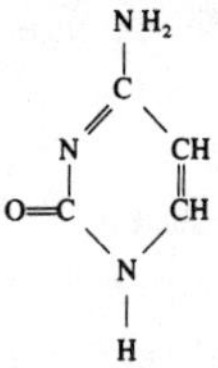

(A) 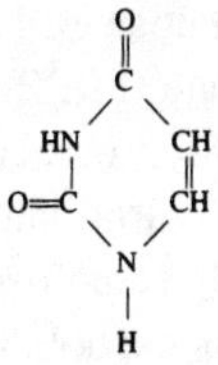 (B) 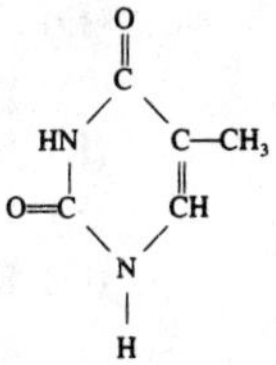 (C) 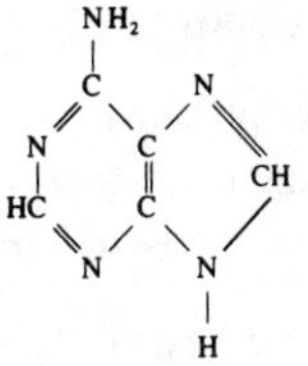 (D) 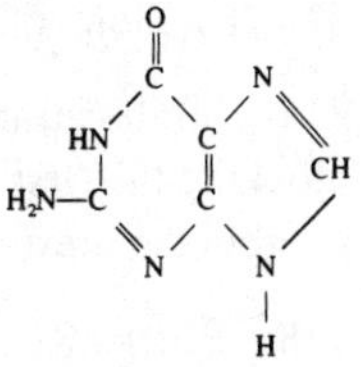 (E)

213. Uracil

214. Thymine

Questions 215–216

Match the sites of attachment below to the correct lettered position on the diagrammatic representation of a eukaryotic gene.

(A)	Regulator
(B)	Promoter
(C)	Operator
(D)	Gene 1
(E)	Gene 2

215. Site of attachment of repressor

216. Site of attachment of CAP protein

Questions 217–220

Match each biochemical action with the appropriate compound.

(A) Streptomycin
(B) Rifamycin B
(C) Tetracycline
(D) Actinomycin D
(E) None of the above

217. Inhibits transcription by blocking the first phosphodiester bond in the presumptive RNA chain

218. Blocks RNA synthesis by binding tightly to double-helical DNA and thereby preventing its use as a template

219. Blocks protein synthesis by inhibiting binding of aminoacyl-tRNA to mRNA by interference at the level of the 30S subunit of the ribosome

220. Decreases the rate of protein synthesis by binding to the 30S subunits of ribosomes and interfering with the reading of mRNA codons

Questions 221–224

Match each description to the correct substance.

(A) Aspartate transcarbamoylase
(B) Allopurinol
(C) Ribose phosphate
(D) 5-Phosphoribosylpyrophosphate (PRPP)
(E) 5-Fluorouracil
(F) Dihydrofolate reductase
(G) Ribose-phosphate pyrophosphokinase

221. Rate-controlling step of pyrimidine synthesis

222. Analogue of hypoxanthine

223. Allosteric inhibition by cytidine triphosphate (CTP)

224. Required substrate for purine biosynthesis

Questions 225–228

For each probe listed below, select its specific target to form the appropriate combination of probe and recombinant technique.

(A) Antibody
(B) mRNA
(C) Products of polymerase chain reaction (PCR)
(D) tRNA
(E) Mutant and normal oligonucleotides
(F) rRNA
(G) cDNA clone

225. cDNA library

226. Products of polymerase chain reaction (PCR)

227. Northern blot

228. Expression library

Questions 229–230

For each enzyme below, select the reaction that it catalyzes.

(A) Initiation of *E. coli* DNA synthesis
(B) Replication of DNA
(C) Repair of DNA lesions
(D) Joining of the ends of DNA
(E) None of the above

229. β-DNA polymerase

230. α-DNA polymerase

Nucleic Acids

Answers

114. The answer is B. *(Stryer, 4/e, pp 120–122.)* A variety of genetic diseases such as sickle cell anemia, Huntington's chorea, and cystic fibrosis can be detected by restriction fragment length polymorphism analysis (RFLP). In order for the RFLP to be able to detect and follow the inheritance of these genes, the detected mutation must be at or closely linked to an altered restriction site. Mutations within the restriction sites change the size of restriction fragments. The different-sized fragments migrate in different positions during electrophoresis of bands visualized by the Southern blot analysis that utilizes ^{32}P-labeled, single-stranded DNA probes.

115. The answer is D. *(Stryer, 4/e, pp 1001–1002.)* All steroid hormones, including the sex hormones estrogen, testosterone and progesterone, act by binding specific cytoplasmic receptors, which then stimulate transcription by binding to specific sites on DNA. These hormones may be contrasted with most nonsteroidal hormones, for example epinephrine, which interact with the cell membrane, causing a second messenger effect. The latter hormones in contrast to steroids act in a matter of minutes, while steroid hormones take hours to have a biologic effect. Recent studies have indicated that specific cytoplasmic receptors for steroid hormones have an extraordinarily high affinity for the hormones. In addition, the receptors contain a DNA-binding region that is rich in amino acid residues that form metal binding fingers. Likewise, thyroid hormone receptors contain DNA-binding domains with metal binding fingers. Like steroid hormones, thyroid hormones are transcriptional enhancers.

116. The answer is C. *(Stryer, 4/e, pp 84–87.)* The melting temperature (T_m) of duplex DNA is the temperature at which half the base pairs are denatured. Adenine-thymine (A-T) base pairs have two hydrogen bonds in contrast to cytosine-guanine (C-G) base pairs, which have three hydrogen bonds. Duplex DNA molecules rich in A-T base pairs have a much lower T_m than those rich in C-G base pairs. As DNA is heated, A-T pairs are the first to melt. The decrease in T_m with increasing C-G base pairs and the converse decrease in T_m with increasing A-T base pairs is a straight-line relationship. According to Chargaff's rules, the content of C + D equals A + G in duplex DNA.

117. The answer is E. *(Stryer, 4/e, pp 91–92, 120–122, 132–136, 794–797, 805–809.)* DNA molecules are normally found in a negatively supercoiled configuration. Supercoiled DNA molecules are more compact than relaxed DNA molecules of the same length and take up less space in the nucleus. The degree of supercoiling is determined by the linking number of DNA. DNA molecules differing only in the linking number are topological isomers (i.e., topoisomers) of each other. Topoisomers of DNA can only be intraconverted by cutting one or both DNA strands. Topoisomerases catalyze the relaxation of supercoiled DNA. Thus, in the case given, circular, supercoiled plasmid DNA topoisomerase catalyzes a three-step process that includes cleavage of one or both strands of DNA, unwinding of a segment of DNA through this break, and sealing of the break. Topoisomerase I cleaves one strand of DNA and type II topoisomerase cleaves both strands. None of the other enzymes listed are capable of relaxing supercoiled DNA. Reverse transcriptase, a retrovirus enzyme that synthesizes a DNA strand complementary to an RNA template, is used to prepare complementary DNA from mRNA that can be expressed in host cells. Restriction endonucleases are capable of cleaving both strands of double helical DNA at specific base sequences. However, they are not capable of resealing the split DNA. The enzyme primase synthesizes the RNA primer that allows DNA synthesis to be carried out. Helicase unwinds duplex DNA ahead of polymerase, synthesizing new DNA at a replication fork. Thus, none of the enzymes other than topoisomerase can cause relaxation of DNA that is otherwise undamaged and unchanged.

118. The answer is C. *(Stryer, 4/e, pp 804–805.)* Since both strands of parental DNA serve as templates for the synthesis of new DNA, it appears that DNA synthesis must be 5′to 3′for one daughter strand and 3′ to 5′ for the other daughter strand at the replication fork. This is contrary to the fact that all DNA polymerases can only synthesize DNA in the 5′ to 3′ direction. This dilemma is solved by understanding that one strand of DNA is synthesized continuously in the 5′ to 3′ direction while the other strand is made up of small fragments known as Okazaki fragments. The small Okazaki fragments are, in fact, synthesized in a 5′ to 3′ direction and then joined together by DNA ligase. Each Okazaki fragment is about 1000 nucleotides long. Thus, while the overall direction of growth of the lagging strand that is made up of small fragments is in fact in the 3′ to 5′ direction, the actual polymerization of individual nucleotides is in the 5′ to 3′ direction.

119. The answer is D. *(Stryer, 4/e, pp 804–810.)* DNA synthesis cannot occur until an RNA primer is made. A specific type of RNA polymerase called *primase* synthesizes a short stretch of RNA of about 5 nucleotides that is complementary to the template DNA strand in duplex DNA near the replication fork. This function cannot be carried out by DNA polymerase. In contrast, both DNA polymerase and RNA polymerase work in the 5′ to 3′ direction and add nucleoside monophosphates from nucleotide triphosphates to the growing polynucleotide chains of DNA or RNA. Only DNA polymerase edits as it synthesizes DNA and fills the gap between Okazaki fragments.

120. The answer is E. *(Stryer, 4/e, pp 894–906.)* Despite some differences, protein synthesis in prokaryotes and eukaryotes is quite similar. The small ribosomal subunit is 30S in prokaryotes and 40S in eukaryotes. The large ribosomal subunit is 50S in prokaryotes and 60S in eukaryotes. The intact ribosome is consequently larger in eukaryotes (80S) and smaller in prokaryotes (70S). At the start of translation, initiation factors, mRNA, and initiation aminoacyl-tRNA bind to the dissociated small ribosomal subunit. The initiation tRNA in prokaryotes is *N*-formyl methionine in prokaryotes and simply methionine in eukaryotes. Only after the small ribosomal subunit is primed with mRNA and initiation aminoacyl-tRNA does the large ribosomal subunit bind to it. Once this happens, elongation factors bring the first aminoacyl-tRNA of the nascent protein to the A site. Then, peptidyl transferase forges a peptide bond between the initiation amino acid and the first amino acid of the forming peptide. The now uncharged initiation tRNA leaves the P site and the peptidyl-tRNA from the A site moves to the now vacant P site with the two amino acids attached. The ribosome advances three bases to read the next codon and the process repeats. When the stop signal is reached after the complete polypeptide has been synthesized, releasing factors bind to the stop signal, causing peptidyl transferase to hydrolyze the bond that joins the polypeptide at the A site to the tRNA. Factors prevent the reassociation of ribosomal subunits in the absence of new initiation complex.

121. The answer is A. *(Stryer, 4/e, pp 949, 964–965.)* Repressors are proteins synthesized by the regulatory gene adjacent to a given operon. Repressors are activated by certain molecules that bind to them and allow them to bind to the operator site of their respective operon, thereby shutting it down. For example, the enzymes needed to synthesize tryptophan are synthesized on the trp operon. One method of regulation of the trp operon is end-product inhibition acting through the trp repressor. When levels of the end product, tryptophan, get too high in the cell, tryptophan complexes with the repressor and binds tightly to the operator. The repressor alone does not.

122. The answer is B. *(Stryer, 4/e, pp 811–813.)* Toxic chemicals, ultraviolet light, and ionizing radiation can damage DNA in a variety of ways.

Phosphodiester bonds in the backbone can be split, opposite strands of the duplex can become cross-linked, and bases can be lost or chemically changed. One of the most common forms of damage to DNA is the production of a pyrimidine dimer by exposure of DNA to ultraviolet light. Adjacent pyrimidine residues on a DNA that have become covalently linked can block replication in gene expression in that area until they are removed. There are two mechanisms of repair. The first involves an enzyme complex encoded by uvrABC genes that detect the pyrimidine dimer. A specific uvrABC enzyme cuts the damaged DNA strands 8 nucleotides away from the dimer on the 5′ side and 4 nucleotides away from the primer on the 3′ side. DNA polymerase I then carries out repair synthesis moving from the 3′ to the 5′ end. Once the newly synthesized DNA is complete, DNA ligase joins the 3′ end to the 5′ end of the original portion of the DNA chain. The other enzymatic mechanism involved in pyrimidine dimer repair is carried out by a photoreactivating enzyme known as DNA photolyase. This form of repair is much more straightforward in that the photolyase is activated by absorption of light in the blue spectral region and simply cleaves the dimer into its original bases. Paradoxically, since both enzymes are active in most cells, it is the cells that are continuously exposed to light that undergo repair to the pyrimidine dimers in a much more rapid fashion.

123. The answer is C. *(Stryer, 4/e, pp 115–116, 864–870.)* Self-splicing of the introns of some primary ribosomal RNA transcripts occurs because of the presence of catalytic RNAs (ribozymes) generated from the introns. This occurs in the absence of protein catalysis. In contrast, the splicing of messenger RNA is carried out by spliceosomes. Spliceosomes are large complexes of three kinds of small ribonucleoprotein particles (snRNPs) and the messenger RNA precursor. The snRNPs are involved in recognizing the 5′ splice site and the 3′splice site and then binding to these sites. Once the spliceosome is bound, it mediates excision of the intron and splicing of the two adjacent exons.

124. The answer is C. *(Stryer, 4/e, pp 982–984.)* Despite the great length of the chromosomes of eukaryotic DNA, the actual replication time is only minutes. This is because eukaryotic DNA is replicated bidirectionally from many points of origin. The hundreds of initiation sites for DNA replication on chromosomes share a consensus sequence called an autonomous replication sequence (ARS). Thus, while eukaryotic DNA, like *E. coli* DNA, is replicated semiconservatively, the use of hundreds of replication forks originating from the autonomous replication sequences allows for a rapid synthesis of chromosomal DNA.

125. The answer is D. *(Stryer, 4/e, pp 115–116, 864–870.)* Until quite recently, it was thought that only proteins could act as biologic catalysts. However, it has since been found that certain types of RNA are capable of undergoing self-splicing. The original discovery was by Dr. Thomas Cech, who found that the intron derived from precursor ribosomal RNA forms a 395-nucleotide RNA molecule that acts as a catalyst for the transformation of other RNA molecules. Such enzymatic RNA molecules are called *ribozymes.* Ribozymes (catalytic RNAs) like this catalyze the cleavage and joining of RNA chains at specific sites without being consumed. Thus, RNA as well as protein can act as a catalyst for enzymatic reactions.

126. The answer is D. *(Stryer, 4/e, pp 60–63, 120–122, 173–174.)* Expression libraries, as opposed to genomic libraries, are often used to screen cDNA clones on the basis of their ability to direct the synthesis of a specific foreign protein of interest in *E. coli.* cDNA molecules are inserted into vectors that allow for efficient expression in bacterial hosts such as *E. coli.* Expression vectors are either plasmids or phages. The vectors are manipulated so that transcription is maximized. This is accomplished by inserting the cDNA into the vector near a small bacterial promoter. In addition, the vectors often encode a ribosome binding site on the mRNA near the initiation codon. As with the detection of a genomic library, spots of bacteria on a replica plate are lysed to release proteins. Replicas of the "master" plate are made by applying sheets of nitrocellulose to the dishes. Radioactive (usually ^{125}I) antibody is added and autoradiography reveals the presence of colonies with the desired protein on the master plate. In this way an immunochemical approach can be used to find the expressed protein. In contrast, a ^{32}P-labeled DNA probe is used to note the presence of a specific DNA sequence in spots on the replica when DNA is being screened. *Southern blotting* refers to the use of the ^{32}P DNA probe of DNA, while *northern blotting* refers to the use of a ^{32}P DNA probe to screen RNA. There is, as yet, no technique named *eastern blotting. Nitrocellulose blotting* simply refers to the transfer of bacteria from a master plate to a nitrocellulose filter for lysing prior to probing,

127. The answer is B. *(Stryer, 4/e, p 814.)* The Ames test is a rapid and relatively inexpensive bacterial assay for determining mutagenicity of potential toxic chemicals. Since many chemical carcinogens are mutagenic, it seems obvious that damage to DNA is a central event in carcinogenesis as well as mutagenesis. Dr. Bruce Ames developed a tester strain of *Salmonella* that has been modified not to grow in the absence of histidine because of a mutation in one of the genes for the biosynthesis of histidine. Toxic chemicals that are mutagens are placed in the center of the plate and result in reversions of the original mutations, so that histidine is synthesized and the mutated revertants multiply in histidine-free media. Dose-response curves of chemicals that are carcinogens are usually linear, indicating a lack of a threshold concentration for mutagenesis and perhaps carcinogenesis. The technique for determining carcinogens can be modified so that mammalian liver homogenate is included in the test system. Many potential carcinogens are formed from noncarcinogenic chemicals that are modified in the liver. Essentially all chemicals known as carcinogens in humans cause mutagenesis in the Ames test.

128. The answer is E. *(Stryer, 4/e, pp 803–809.)* Before DNA replication can actually begin, unwinding protein must open segments along the DNA double helix. DNA-directed RNA polymerase (primase) catalyzes the synthesis of a complementary RNA primer of approximately 50 to 100 bases on each DNA strand. Then DNA-directed DNA polymerase III adds deoxyribonucleotides to the 3′ end of the primer RNA, which replicates a segment of DNA, the Okazaki fragment. DNA polymerase I then removes the primer RNA and adds deoxyribonucleotides to fill the gaps between adjacent Okazaki fragments. The fragments are finally joined together by DNA ligase to create a continuous DNA chain.

129. The answer is B. *(Stryer, 4/e, pp 788–791.)* In the classic double helical model of DNA proposed by Watson and Crick, the purine and pyrimidine bases attached to the sugar backbone are perpendicular to the axis and parallel to each other. They are paired and held together by hydrogen bonds. Each strand composing the double helix is different and antiparallel. The 3′ end of one strand is opposite the 5′ end of its complement and vice versa. It is this complementary nature of DNA that allows the strands to be templates for one another during DNA replication.

130. The answer is B. *(Stryer, 4/e, pp 721–723.)* *S*-Adenosylmethionine donates the methyl group of methionine in many biochemical reactions, which leaves *S*-adenosylhomocysteine. These reactions include methylation of guanidinoacetic acid to yield creatine and of phosphatidylethanolamine to yield phosphatidylcholine. *S*-Adenosylmethionine is regenerated from *S*-adenosylhomocysteine in the activated methyl cycle.

131. The answer is C. *(Stryer, 4/e, pp 902–903.)* In a general sense, the mechanism of protein synthesis in eukaryotic cells is similar to that found in prokaryotes; however, there are significant differences. Cycloheximide inhibits elongation of proteins in eukaryotes, while erythromycin causes the same effect in prokaryotes. Thus, one is an antibiotic beneficial to humans, while the other is a poison. Cytoplasmic ribosomes of eukaryotes are larger, sedimenting at 80S instead of 70S. While eukaryotic cells utilize a specific tRNA for initiation, it is not formylated as in bacteria. Finally, eukaryotic mRNA always specifies only one polypeptide as opposed to prokaryotic mRNA, which may specify the synthesis of more than one gene product per mRNA.

132. The answer is A. *(Stryer, 4/e, pp 75–77.)* The sugars that are the backbone of nucleic acids are pentoses. Either D-ribose, which contains hydroxyl groups on each carbon, or 2-deoxy-D-ribose, which has no hydroxyl on the C2 carbon, is used in the synthesis of RNA and DNA. Both sugars occur in the furanose form in nucleic acids. In this form, the C4 hydroxyl is joined to the C1 carbon to form a five-membered ring. In nucleosides, the C1 carbon is linked to the N1 of a pyrimidine or the N9 of a purine base by a β-N-glycosidic linkage.

133. The answer is E. *(Stryer, 4/e, pp 75–77.)* A nucleotide is composed of a nucleoside esterified to a phosphate at C5 of the pentose sugar. A nucleoside is a pyrimidine or purine base linked to a pentose sugar. The precursors of DNA are deoxyribonucleotides, while the monomeric units composing RNA are ribonucleotides. Each name is dependent upon whether a ribose or deoxyribose sugar is present.

134. The answer is B. *(Stryer, 4/e, pp 842–845.)* Promoter sites are initiation sites for transcription. Transcription starts when RNA polymerase binds to the promoter. It then unwinds the closed promoter complex where DNA is in the form of a double helix to form the open promoter complex in which about 17 base pairs of template DNA are unwound. RNA synthesis then begins with either a pppA or a pppG inserted at the beginning 5′-terminus of the new RNA chain, which is synthesized in the $5' \rightarrow 3'$ direction.

135. The answer is B. *(Stryer, 4/e, pp 853–854.) Amanita phalloides,* the mushroom known as the destroying angel, leads to a considerable number of deaths each year. It produces the cyclic octapeptide amanitin, which is composed of several unusual amino acids. The major site of the toxin's action is RNA polymerase II, to which it binds very tightly. Polymerase II is responsible for the synthesis of mRNA.

136. The answer is E. *(Stryer, 4/e, pp 80–83.)* The single-stranded ends of the duplex DNA shown are complementary; that is, they repeat in the same sequence on the right and left sides. Since there is terminal repetition, the ends are cohesive. Thus, the polynucleotide could bend back on itself and hybridize to form a double-stranded circle. Phosphodiester bonds on each strand would have to be formed to close each strand. The single-stranded ends could possibly have been formed from the treatment of a duplex piece of DNA with a 5′-exonuclease. Intact plasmid DNA is present in double-stranded circles, which is different from the polynucleotide drawn. Neither one of the terminal ends is a palindrome, which is a symmetric sequence that reads the same from either end.

137. The answer is D. *(Stryer, 4/e, pp 84–87.)* The melting temperature (T_m) of duplex DNA is the temperature at which half the base pairs are denatured. Because cytosine-guanine (C-G) base pairs have three rather than two hydrogen bonds, a high content of C + G increases T_m by virtue of the positive straight-line relationship. According to Chargaff's rules, in duplex DNA the content of A + G = C + T.

138. The answer is A. *(Stryer, 4/e, pp 88–89.)* DNA replication entails pairing of thymine with adenine and guanine with cytosine. The chains of the double helix are thus bonded in part by a hydrogen linkage between amino and keto groups. The strands themselves are synthesized in an antiparallel direction, i.e., the $5' \rightarrow 3'$ sequence of phosphodiester bridges mentioned in the question specifies its complement in a $3' \rightarrow 5'$ direction.

139. The answer is E. *(Stryer, 4/e, pp 109–110.)* AUG is the codon for both methionine and *N*-formylmethionine. In *E. coli,* AUG is the chain-initiating codon and *N*-formylmethionine is the first amino acid incorporated into the nascent polypeptide. In mammals, AUG is also thought to be the codon for chain initiation, but methionine (rather than *N*-formylmethionine) is the N-terminal amino acid.

140. The answer is C. *(Stryer, 4/e, pp 912–915.)* By using recombinant DNA techniques, mRNAs can be produced that yield chimeric proteins. By forming mRNAs that produce otherwise cytosolic proteins like α-globin with a cleavable amino terminal signal sequence, it has been shown that this sequence contains all the information needed to direct the translocation of protein across endoplasmic reticulum; that is, an otherwise cytosolic protein will now become a secretory protein and be translocated into the lumen of endoplasmic reticulum. This experiment is carried out by adding chimeric mRNA to an in vitro system of protein synthesis composed of endoplasmic reticulum vesicles, ribosomes, tRNAs, and other factors required for protein synthesis. Without the modified amino terminal signal sequence, the α-globin is released into the experimental solution and with the signal sequence it is synthesized into the lumen of the endoplasmic reticulum vesicles.

141. The answer is E. *(Stryer, 4/e, pp 808–811.)* Insertion of one extra nucleotide causes a frameshift mutation and mistranslation of all the mRNA transcribed from beyond that point in the DNA. All the other mutations cited in the question usually cause an error in the identity of only one amino acid (A or B), or removal of one amino acid from the sequence D, or no error at all in the amino acid sequence C. There is a chance that either A or B will give a "nonsense," or chain-terminator, mutation, and this is about as likely to be lethal as is a frameshift.

142. The answer is A. *(Stryer, 4/e, pp 83–84.)* Semiconservative double-stranded DNA replication demands that from each half (strand) of the labeled parental DNA, a complement will be synthesized that maintains the parental structure intact. Therefore, the first round of replication in a cold (unlabeled) solution will yield two molecules that are half-labeled. The second round will yield two half-labeled and two unlabeled molecules of double-stranded DNA.

143. The answer is C. *(Stryer, 4/e, pp 879–880, 886, 894–901.)* ATP is required for the esterification of amino acids to their corresponding tRNAs. This reaction is catalyzed by the class of enzymes known as *aminoacyl-tRNA synthetases.* Each one of these enzymes is specific for one tRNA and its corresponding amino acid.

$$\text{Amino acid} + \text{tRNA} + \text{ATP} \rightarrow \text{Aminoacyl-tRNA} + \text{AMP} + \text{PP}_i$$

As with most ATP hydrolysis reactions that release pyrophosphate, pyrophosphatase quickly hydrolyzes the product to P_i, which makes the reaction essentially irreversible. Since ATP is hydrolyzed to AMP and PP_i during the reaction, by convention, the equivalent of two high-energy phosphate bonds is utilized.

144. The answer is E. *(Stryer, 4/e, pp 170–171.)* In Hb S, a valine residue replaces a glutamic acid on the β-chain as the result of a point mutation in one nucleotide base. This single nucleotide alteration at the second position of the triplet consists of a change of thymine to adenine.

145. The answer is B. *(Stryer, 4/e, pp 901–902.)* During the course of protein synthesis on a ribosome, peptidyl transferase catalyzes the formation of peptide bonds. However, when a stop codon is reached, such as UAA, UGA, or UAG, aminoacyl-tRNA does not bind to the A site of a ribosome. One of the proteins known as a release factor binds to the specific trinucleotide sequence present. This binding of the release factor activates peptidyl transferase to hydrolyze the bond between the polypeptide and the tRNA occupying the P site. Thus, instead of forming a peptide bond, peptidyl transferase catalyzes the hydrolytic step that leads to the release of newly synthesized proteins. Following release of the polypeptide, the ribosome dissociates into its major subunits.

146. The answer is D. *(Stryer, 4/e, pp 876–878.)* The rule that uridine replaces thymine in RNA is correct with one minor exception. All tRNAs contain one thymine linked to a ribose sugar. In addition, tRNAs contain a variety of other unusual bases. These are mostly methylated uridine, adenine, guanine, and cytosine derivatives formed by posttranslational modification. Inosine is also present in tRNAs.

147. The answer is A. *(Stryer, 4/e, pp 844–845.)* Sigma factor is the subunit of RNA polymerase that confers specificity of initiation on the core enzyme. In the presence of sigma factor, RNA polymerase will choose the correct strand of duplex DNA for transcription and will initiate transcription at the appropriate promoter region.

148. The answer is D. *(Stryer, 4/e, pp 915–918.)* The directing of nascent polypeptide chains to endoplasmic reticulum is regulated by signal recognition particles (SRPs). The signal sequence of a nascent protein is recognized by SRP, which complexes with the ribosome, mRNA, and the nascent protein. The complexed SRP then binds to an SRP receptor on the surface of the endoplasmic reticulum. After the ribosome is transferred to ribophorins and the translocation begins, SRP is released back into the cytosol. Ribosomes with nascent protein without a signal sequence do not participate in this process and, instead, synthesize proteins that are released into the cytosol.

149. The answer is A. *(Stryer, 4/e, pp 112–115.)* The differences between prokaryotic and eukaryotic mRNA are profound. In bacteria, primary transcripts of mRNA can be used immediately and directly for translation, while in eukaryotes, primary mRNA must be processed, modified by splicing enzyme, and transported from the nucleus to the cytoplasm. Prokaryotic mRNA can either code for a single polypeptide (monocistronic) or for several polypeptide chains (polycistronic). Eukaryotic mRNA is always monocistronic. Primary transcripts of eukaryotic mRNA are synthesized with introns that must be removed. In addition, eukaryotic mRNAs contain 5′ caps of 7-methylguanosine and often have 3′ poly A tails.

150. The answer is D. *(Stryer, 4/e, pp 128, 902–903.)* Tetracycline blocks the binding of aminoacyl-tRNA with initiator sites on 30S subunits of ribosomes, which causes an irreversible inhibition of protein synthesis. The transcription of DNA, the binding of mRNA to ribosomes, and the formation of aminoacyl-tRNAs are not affected. Release of peptides from mRNA-tRNA complexes does not occur inasmuch as peptide bonds are not formed.

151. The answer is E. *(Stryer, 4/e, pp 902–903.)* Chloramphenicol interacts with the 50S subunit of prokaryotic ribosomes to inhibit the process of polypeptide chain elongation. Streptomycin, on the other hand, binds to the 30S subunit of ribosomes, which results in decreased rates of protein synthesis, but additionally it induces misreading of mRNA codons.

152. The answer is B. *(Stryer, 4/e, pp 886–889.)* Recognition of codons on mRNA by the anticodons of tRNAs does not obey strict Watson-Crick pairing rules. Some tRNAs will recognize and bind to more than one codon. This is because the base pairing at the third position (3′ end) of the codon is not stringent as in the first two positions. Certain bases of the 5′ position of the anticodon may pair with more than one base at the 3′ position of the codon. Thus, the "wobble," or imprecision, in the pairing of anticodon to codon explains part of the degeneration of the genetic code.

153. The answer is B. *(Stryer, 4/e, pp 911–914.)* The two subunits of ribosomes are composed of proteins and rRNA. Ribosomes are found in the cytoplasm, in mitochondria, and bound to the endoplasmic reticulum. Transcription refers to the synthesis of RNA complementary to a DNA template and has nothing immediately to do with ribosomes.

154. The answer is E. *(Stryer, 4/e, pp 819–839.)* The process of transduction involves the transfer of a portion of DNA from one bacterium to the chromosome of another bacterium by means of a viral infection. Conjugation is the transfer of a so-called male chromosomal DNA to the DNA of an acceptor, or female, bacterial cell. Colinearity defines the relationship between genes and proteins in that the sequence of amino acids in proteins is a result of the sequence of base triplets in template genes. Recombination is simply the exchange of sequences between two molecules of DNA. Transformation results when exogenous DNA fragments are incorporated into the chromosome of another organism.

155. The answer is A. *(Stryer, 4/e, pp 811–812.)* UV irradiation causes thymine dimers to form in DNA. Replication is inhibited in cells until the pyrimidine dimers are removed. Removal of the damaged areas occurs in two ways. The process can be simply reversed by a photoreactivating enzyme that cleaves the dimers and yields the original bases. Blue light is required. Alternatively, the dimer is removed. A UV-specific excinuclease nicks the dimer on its 5′ side. DNA polymerase I replicates the damaged sequence, while the damaged sequence swings out. Finally, the damaged piece is hydrolyzed by the 5′→ 3′ exonuclease activity of the DNA polymerase I. DNA ligase joins the new piece to the original DNA at the cleavage site.

156. The answer is D. *(Stryer, 4/e, pp 858–859.)* The primary transcripts of all eukaryotic mRNAs are capped at their 5′ end. Prokaryotic RNAs and eukaryotic tRNA and rRNA are not capped. The cap is composed of 7-methylguanylate attached by a pyrophosphate linkage to the 5′ end. This is known as cap 0. One of the adjacent riboses is methylated in cap 1, and both of the adjacent riboses are methylated in cap 2. The cap protects the 5′ ends of mRNAs from nucleases and phosphatases and is essential for the recognition of eukaryotic mRNAs in the protein-synthesizing system. When prokaryotic monocistronic mRNAs are artificially capped, translation will occur in a eukaryotic, in vitro translation system.

157. The answer is A. *(Stryer, 4/e, pp 964–965.)* RNA polymerase synthesizes RNA from a DNA template in a 5′ → 3′ direction. Since the antiparallel strand of DNA is utilized, it follows that the upper strand in the example given was utilized. In contrast to DNA synthesis, RNA polymerase requires no primer.

158. The answer is A. *(Stryer, 4/e, pp 636–637, 748.)* The first control step in the synthesis of pyrimidines is the condensation of carbamoyl phosphate and aspartate to form *N*-carbamoylaspartate. The cytosolic carbamoyl phosphate synthetase involved in producing the carbamoyl phosphate used is inhibitable by uridine monophosphate. It has a low activity in the liver of mammals and is distinctly separate from the mitochondrial carbamoyl phosphate synthetase that is involved in urea biosynthesis. The latter enzyme has a high activity and is activated by acetylglutamate.

159. The answer is C. *(Stryer, 4/e, pp 896–902.)* Two molecules of GTP are used in the formation of each peptide bond on the ribosome. In the elongation cycle, binding of aminoacyl-tRNA delivered by EF-Tu to the A site requires hydrolysis of one GTP. Peptide bond formation then occurs. Translocation of the nascent peptide chain on tRNA to the P site requires hydrolysis of a second GTP. The activation of amino acids with aminoacyl-tRNA synthetase requires hydrolysis of ATP to AMP plus PP_i.

160. The answer is D. *(Stryer, 4/e, pp 848–849.)* DNA contains stop signals, some of which require rho protein. This has been demonstrated by examining the synthesis of mRNA in the presence and absence of rho protein. In the absence of rho protein, longer RNA molecules are often synthesized. This would seem to indicate that mRNA length can be controlled by the cell. In addition, antiterminator proteins are needed to allow certain genes to be properly expressed.

161. The answer is B. *(Stryer, 4/e, pp 102, 855–856.)* Genes encoding proteins contain promoter sites that bind RNA polymerase and determine where transcription begins. Promoter sites are upstream (on the 5′side) of the first nucleotide to be transcribed in the 1 position. In bacteria, two sequences are promoter sites. The Pribnow box (TATAAT) is ten nucleotides upstream (−10). The other promoter site (TTGACA) is −35. In eukaryotic genes, up to three sites may exist. Nearly all eukaryotic genes contain the TATA box (also called the *Hogness box*) with the TATAAA consensus sequence at about −25. Many also contain a CAAT site and a GC box at variable positions between −40 and −110 upstream.

162. The answer is B. *(Stryer, 4/e, pp 879–902.)* Amino acids are activated by specific aminoacyl-tRNA synthetases that join them to their unique tRNA with the consumption of two high-energy phosphate bonds. Next, the 30S rRNA subunit binds initiation factor 3, mRNA, and initiation factor 1. Aminoacyl-tRNA, initiation factor 2, and GTP join the 30S subunit complex, which unites with the 50S subunit. Aminoacyl-tRNA is bound to the initiation codon of mRNA, a process driven by the hydrolysis of GTP to guanosine diphosphate (GDP) and inorganic orthophosphate (P_i). The three initiation proteins are then released from the complex. In ensuing steps of elongation, aminoacyl-tRNA binds to elongation-factor protein together with GTP. GTP hydrolysis energizes the addition of the new aminoacyl-tRNA to the active (A) site of the ribosome. Next, peptidyl transferase catalyzes formation of a peptide bond between the carboxyl group of the preceding amino acid and the amino group of the newly added amino acid. The ribosome advances one codon forward on the mRNA and moves the peptidyl-tRNA from the A site to the peptidyl (P) site, thereby clearing the A site for the next incoming aminoacyl-tRNA. This final step, which is called *translocation,* requires a second elongation factor and consumes another mole of GTP.

163. The answer is D. *(Stryer, 4/e, p 906.)* The gene that produces the deadly toxin of *Corynebacterium diphtheriae* comes from a lysogenic phage that grows in the bacteria. Prior to immunization, diphtheria was the primary cause of death in children. The protein toxin produced by this bacterium inhibits protein synthesis by inactivating elongation factor 2 (EF2, or translocase). Diphtheria toxin is a single protein composed of two portions (A and B). The B portion enables the A portion to translocate across a cell membrane into the cytoplasm. The A portion catalyzes the transfer of the adenosine diphosphate ribose unit of NAD^+ to a nitrogen atom of the diphthamide ring of EF2, thereby blocking translocation. Diphthamide is an unusual amino acid residue of EF2.

164. The answer is D. *(Stryer, 4/e, pp 80–88.)* Like bacterial DNA, eukaryotic DNA is replicated in a semiconservative manner. However, in contrast to most bacterial DNA, which is circular in structure, nuclear chromosomal DNA is a single, uninterrupted molecule that is linear and unbranched. A eukaryotic chromosome contains a strand of DNA at least 100 times as large as the DNA molecules found in prokaryotes. Eukaryotic, but not prokaryotic, DNA molecules are bound to small basic proteins called *histones.* The histone-DNA complex formed is referred to as *chromatin.*

165. The answer is E. *(Stryer, 4/e, pp 812–813.)* Xeroderma pigmentosum appears to be due to the inability of an excision-repair system to remove thymine dimers, which are formed on exposure of DNA to ultraviolet radiation. This results in a deficiency of the ability to repair the damaged DNA. Mutagenesis by this mechanism is presumably the basis for the multiple neoplasms that occur in patients who have this disease.

166. The answer is E. *(Stryer, 4/e, pp 902–903.)* Puromycin is a structural analogue of the aminoacyl end of the tRNA. It irreversibly reacts with the peptidyl-tRNA, thereby terminating protein synthesis. Streptomycin, like tetracycline and chloramphenicol, inhibits ribosomal activity. Mitomycin covalently cross-links DNA, which prevents cell replication. Rifampicin is an inhibitor of DNA-dependent RNA polymerase.

167. The answer is B. *(Stryer, 4/e, pp 992–993.)* The nucleolus is an organelle unique to eukaryotic cells. It is the site where hundreds of copies of genes repeated in tandem for three of the four ribosomal RNAs are transcribed by RNA polymerase I to give a 45S primary transcript. Enzymatic modification and cleavage remove spacer regions to yield 28S, 18S, and 5.8S ribosomal RNA. The 5S subunit is synthesized by RNA polymerase III in the nucleoplasm, rather than in the nucleolus. Ribosomal proteins combine with the ribosomal subunits to assemble into a 60S subunit containing the 5S, 5.8S, and 28S RNAs and a 40S subunit containing the 18S RNA. Combined, the two subunits produce a functional eukaryotic ribosome with a sedimentation coefficient of 80S.

168. The answer is A. *(Stryer, 4/e, pp 811–812.)* Ultraviolet radiation (260 nm) causes dimerization between pyrimidines on the same polynucleotide strand. Replication will stop at that point. The dimers can be repaired by a mechanism that involves three enzymes. First, a highly specific exonuclease-protein complex detects the structural distortion produced by the dimer and cuts the DNA strand 8 nucleotides away from the dimer on the 5′ side and 4 nucleotides away on the 3′ side. DNA polymerase I carries out repair synthesis of the 12-base sequence. Finally, DNA ligase joins the 3′ end of the new piece to the original strand.

169. The answer is D. *(Stryer, 4/e, pp 913–915.)* The signal hypothesis attempts to explain the biosynthesis of secretory and membrane proteins. Since all eukaryotic ribosomes are intrinsically the same, whether or not a particular ribosome attaches to endoplasmic reticulum depends on the synthesis of a single peptide from 15 to 30 amino acid residues near the amino end of a nascent polypeptide chain. Most secretory and membrane proteins are synthesized continuously with such a signal peptide that binds the ribosome to a receptor site on the endoplasmic reticulum. The protein passes through the membrane into the lumen during its synthesis, and the signal peptide is excised by a specific peptidase on the luminal surface of the endoplasmic reticulum. The attachment of ribosomes to mRNA, either singly or in multiples on long messengers (polyribosomes), and the release of proteins from ribosomes are unchanged for secretory protein synthesis.

170. The answer is D. *(Stryer, 4/e, p 811.)* 5-Bromouracil is incorporated into DNA, in place of thymidine, to yield a denser DNA. The newly synthesized DNA fragments can then be quantitated by centrifugation through density gradients of cesium chloride. 5-Bromouracil is neither more reactive nor more sensitive to cleavage than thymidine, nor does it cause frameshift mutations, as do the acridine dyes.

171. The answer is C. *(Stryer, 4/e, pp 120–121, 792–793.)* A palindromic sequence in base pairs of double-stranded DNA demonstrates twofold rotational symmetry; that is, the strand reads the same from 3′ to 5′ as its complement reads from 5′ to 3′. Many restriction enzymes recognize specific palindromic sequences of from four to six base pairs and hydrolyze the phosphodiester bonds central to the symmetric axis in each strand. The cleavage site of the given palindrome is

G-T-C-A-T-G-A-C
×
C-A-G-T-A-C-T-G

172. The answer is B. *(Stryer, 4/e, pp 109–110, 175.)* The replacement of the codon UAG with UAA would be a silent mutation since both codons are "stop" signals. Thus, transcription would cease when either triplet was reached. There are three termination codons in mRNA: UAG, UAA, and UGA. These are the only codons that do not specify an amino acid. A missense or a substitution mutation is the converting of a codon specifying one amino acid to another codon specifying a different amino acid. A nonsense mutation converts an amino acid codon to a termination codon. A suppression counteracts the effects of another mutation at another codon. The addition or deletion of nucleotides results in a frameshift mutation.

173. The answer is C. *(Stryer, 4/e, pp 75–77.)* The pyrimidine thymine and the sugar deoxyribose are incorporated into DNA, not RNA, although thymine can exist as a ribonucleoside in tRNA. Adenine, cytosine, and guanine are incorporated into both RNA and DNA. Ribose sugar and the base uracil are specific to all the forms of RNA.

174. The answer is B. *(Stryer, 4/e, pp 112–113, 836–839.)* All genes studied to date in prokaryotes are single-copy and produce a continuous strand of complete mRNA. In contrast, transcription of RNA in eukaryotes is quite diverse and complicated. Only some of the genes coding for proteins are continuous. These include some tRNA genes and genes coding for rRNA and histone mRNA. However, most genes coding for proteins in eukaryotes are discontinuous or split; that is, intervening sequences (introns) of DNA are transcribed together with mRNA-coding sequences (exons) to produce an immature RNA. The introns transcribed must be removed in the formation of mature mRNA. For example, in mouse immunoglobulin H chain, four introns must be removed from the newly synthesized mRNA. While many major proteins are encoded by single-copy genes in eukaryotes, other genes are repeated in tandem many times (e.g., rRNA and histones). Thus, single-copy genes, genes repeated in tandem, and split genes all exist in eukaryotic cells.

175. The answer is B. *(Stryer, 4/e, pp 120–121.)* *Kilobase* refers to a unit of length equivalent to either 1000 bases of a single-stranded DNA molecule or 1000 base pairs of a double-stranded molecule. A kilobase (kb) of double-stranded DNA corresponds to a mass of approximately 660 kilodaltons and a contour length of 0.34 μm. Single-stranded DNA has about the same length and about half the weight. Viral DNA ranges from 5 to 200 kb, prokaryotic DNA from 760 kb for mycoplasma to 4000 kb for *E. coli,* and eukaryotic DNA from 13,000 kb for yeast to about 3,000,000 kb for man.

176. The answer is B. *(Stryer, 4/e, pp 132–136.)* The formation of complementary DNA (cDNA) requires the appropriate mRNA. Reverse transcriptase from RNA retroviruses is the key enzyme in the synthesis of cDNA from mRNA. Reverse transcriptase will work if an oligo dT primer with a free 3′-OH end is provided to pair with the poly-A sequence found at the 3′ end of most eukaryotic mRNAs. Following cDNA synthesis, the mRNA template can be removed by alkali digestion, which leaves the cDNA free for insertion into plasmid or phage vectors.

177. The answer is C. *(Stryer, 4/e, pp 109–111, 886–888.)* Many messenger RNA molecules code for more than one polypeptide chain; this is often the basis for coordinate control of enzyme synthesis in bacteria. There are no "spacers" to mark the end of one codon and the beginning of another. There are three "nonsense" codons. According to the wobble hypothesis of Crick, certain anticodons can pair with more than one codon where variability occurs at the 3' position of the codon. Genes and encoded protein products are colinear as determined by genetic mapping.

178. The answer is B. *(Stryer, 4/e, pp 976–980.)* Eukaryotic DNA is found on chromosomes, each of which is composed of a single linear molecule of double-helical DNA. Approximately 6000 replication forks per DNA molecule are observed during replication. Replication occurs bidirectionally from any initiation points. DNA polymerase α, which contains no nuclease activity, plays the major role in chromosome replication. Many repeated base sequences are observed in eukaryotic DNA; more than 30 percent of human DNA is repetitive. Centromeric DNA, DNA coding for ribosomal RNA, and DNA coding for histone messenger RNA are highly repetitive.

179. The answer is C. *(Stryer, 4/e, pp 174–175, 809–810.)* Point mutations that are frameshift mutations put the normal reading-frame out of register by one base pair. The insertion of an extra base pair or the deletion of one or more base pairs falls into this category. Transitions and transversions are not frameshift mutations; they are substitutions of one base pair for another. Substitutions are the most common type of mutation. In transitions, a purine is replaced by a purine or a pyrimidine by a pyrimidine. In transversions, a purine is replaced by a pyrimidine or vice versa. It has been suggested that transitions occur spontaneously owing to the tautomeric changes in base-hydrogen-bond locations. Transversions can be caused by defective DNA polymerases.

180. The answer is A. *(Stryer, 4/e, pp 740–746.)* During purine ring biosynthesis, the amino acid glycine is completely incorporated to provide C-4, C-5, and N-7. Glutamine contributes N-3 and N-9, aspartate provides N-1, and derivatives of tetrahydrofolate furnish C-2 and C-8. Carbon dioxide is the source of C-6. In pyrimidine ring synthesis, C-2 and N-3 are derived from carbamoyl phosphate, while N-1, C-4, C-5, and C-6 come from aspartate.

181. The answer is C. *(Stryer, 4/e, pp 75–77.)* Since a chain of DNA is polar, with one end having a 5′-OH group and the other end a 3′-OH group, a conventional means for simply communicating linear structure is necessary. Convention dictates that base sequences are written in a 5′ to 3′ direction, much like amino acids are written in an amino end to carboxyl end direction. Thus, the complement to 5′-AGCT-3′ is 5′-AGCT-3′ when written conventionally and 3′-TCGA-5′ when written in a truly complementary way.

182. The answer is A. *(Stryer, 4/e, pp 120–122, 854–856, 950–951, 977–982.)* In eukaryotes, RNA polymerase binds to promoter sites upstream from the start site. These include the TATA box (TATAAT), the CAAT box, and the GC box. Repressor protein, which is produced by the regulator gene of an operon, binds to a specific sequence of the operator gene. Inducer protein forms a complex with repressor and prevents the latter from interacting with the operator gene. Restriction endonucleases recognize specific sequences in double-helical DNA and cleave both strands. Histones nonspecifically bind to chromosomal DNA and constitute about half the mass of eukaryotic chromosomes.

183. The answer is C. *(Stryer, 4/e, pp 888–889, 904.)* Prokaryotic ribosomes have a sedimentation coefficient of 70S and are composed of 50S and 30S subunits. Eukaryotic cytoplasmic ribosomes, either free or bound to endoplasmic reticulum, are larger—60S and 40S subunits that associate to an 80S ribosome. Nuclei ribosomes are attached to the endoplasmic reticulum of the nuclear membrane. Ribosomes in chloroplasts and mitochondria of eukaryotic cells are more similar to prokaryotic ribosomes than to eukaryotic cytosolic ribosomes. Like bacterial ribosomes, chloroplast and mitochondrial ribosomes use a formylated tRNA. In addition they are sensitive to many of the inhibitors of protein synthesis in bacteria.

184. The answer is D. *(Stryer, 4/e, pp 951–952.)* The operon, as proposed by Jacob and Monod, can be described as a group of functionally related structural genes that map close to each other in the chromosome and can be turned on or off together through the same regulatory loci. The DNA elements of the model are the regulator, the promoter, the operator, and the structural genes. The regulator codes for the synthesis of a repressor protein that binds the operator and thereby prevents attachment of RNA polymerase to the promoter and initiation of synthesis. Inducers (such as isopropylthiogalactoside in the case of the lac operon) bind and inactivate the repressor, which allows synthesis to proceed. Histones are proteins that are associated with DNA. Retroviruses are RNA-containing viruses that synthesize double-helix DNA with a reverse transcriptase.

185. The answer is D. *(Stryer, 4/e, pp 80–83.)* Double-stranded DNA is arranged in a double helix originally deduced by Watson and Crick. The double helical structure of duplex DNA is different than the α-helical structure of portions of proteins. The α-helical structure of proteins is formed of one chain of proteins stabilized by individual hydrogen bonds between components of the amide bonds, that is, between the carbonyl oxygens and the amide nitrogens. In contrast, the hydrogen bonding in double-stranded DNA is important to allow each strand to act as a template for the other complementary strand with adenine bonding to thiamine and cytosine bonding to guanine. Hydrophobic stacking between bases in the hydrophobic interior of the double strand actually makes a greater contribution to the stability of the DNA double helix than hydrogen bonding. DNA and protein helices are both composed of polymers of subunits (amino acids and nucleotides) held together by a covalently linked backbone. As pointed out above, hydrogen bonding is important to both the double helix of DNA and the α-helix of proteins. Finally, both are, in fact, spiral structures, although only the helix of proteins is an α-helix.

186. The answer is E. *(Stryer, 4/e, pp 100–102, 845–849.)* DNA and RNA synthesis mechanisms are similar in a variety of ways. First, both involve a hydrophilic attack on the 5′ triphosphate of the incoming nucleotide by the 3′ hydroxyl group of the last nucleotide of the elongating chain. Pyrophosphate is released from the 5′ triphosphate. Second, the polymerases for both DNA and RNA require activated nucleotide precursors and magnesium ion in order to be functional. Third, DNA or RNA polymerase chooses the correct incoming base to incorporate by using base-pairing information of the strand being copied as well as the geometry of the incoming base. Fourth, in both RNA synthesis and DNA replication, the direction is 5′ $\rightarrow$ 3′. One of the major differences between RNA synthesis and DNA synthesis is that RNA polymerases can initiate an RNA chain without primer. DNA synthesis is initiated with a primer.

187. The answer is D. *(Stryer, 4/e, pp 135–136, 808–809, 811–812, 834–837.)* Oncogenes are cancer-producing genes. They are closely related to normal cellular genes and are often tyrosine kinases, growth factors, or receptors for growth factors. The expression of oncogenes leads to the translation and eventual transcription of the protein product of the oncogene. Thus, DNA synthesis does not occur during the expression of oncogenes. In contrast, template-directed DNA synthesis does occur during the repair of thymine dimers, the polymerase chain reaction, the functioning of the replication fork, and growth of RNA tumor viruses. In the final stages of the repair of thymine dimers, once the dimer has been excised, DNA polymerase I enters the gap to carry out template-directed synthesis. In functioning of the replication fork, DNA polymerase III holoenzyme carries out synthesis of DNA during replication. Template-directed synthesis is required for the growth of RNA tumor viruses (retroviruses). Once released into the host cytoplasm, retroviral RNA synthesizes both the positive and minus strands of DNA, using reverse transcriptase. This unique enzyme catalyzes the initial RNA-directed DNA synthesis, hydrolysis of RNA, and then DNA-directed DNA synthesis. The newly formed viral DNA duplex integrates into the host cell DNA prior to transcription. In this form, the retrovirus is inherited by daughter host cells. The polymerase chain reaction is a method of amplifying the amount of DNA in a sample or enriching particular DNA sequences in a population of DNA molecules. In the polymerase chain reaction, oligonucleotides complementary to the ends of the desired DNA sequence are used as primer for multiple rounds of template-directed DNA synthesis.

188. The answer is E. *(Stryer, 4/e, pp 126–128, 793–794, 808–810.)* In the leading strand, DNA is synthesized continuously in the 5′ to 3′ direction by DNA polymerase. In contrast, in the lagging strand, which is in the 3′ to 5′ direction, DNA polymerase III synthesizes small (approximately 1000 nucleotides) Okazaki fragments. For the synthesis of these small fragments, all the same roles and steps apply except that additional enzymes are needed to fill the gap between the fragments and join the fragments. Consequently, DNA ligase is repeatedly needed to join the ends of the DNA fragments along the growing lagging strand. DNA ligase catalyzes the formation of a phosphodiester bond between the 3′ hydroxyl group at the end of one DNA chain and the 5′ DNA phosphate group at the end of the other. DNA ligase is only functional when double-helical DNA molecules are the substrate. It does not work on single-stranded DNA. DNA ligase effects the joining of strands of DNA not only during the normal synthesis of DNA, but during the splicing of DNA chains in genetic recombination as well as the repair of damaged DNA.

189. The answer is D. *(Stryer, 4/e, pp 876–888.)* While transfer (t) RNA molecules have many features in common, the primary feature that sets them apart is their specificity for different amino acids and the corresponding specific differences of their anticodons. Each tRNA is an L-shaped, single chain composed of up to 93 ribonucleotides. Each contains up to 15 methylated bases and about half of the nucleotides are base-paired into double helices. The 5′ end is often guanosine and is always phosphorylated. The 3′ +end is CCA. Activated amino acids attach to the terminal 3′-hydroxyl group of the adenosine.

190. The answer is C. *(Stryer, 4/e, pp 75–83.)* Functional nuclear DNA is found in a double helix with each strand complementary to the other. All the following properties derive from this structural association. Adenine binds exclusively with thymine using two hydrogen bonds per pair, while guanine binds only with cytosine using three hydrogen bonds. Thus, the ratio of thymine + cytosine/guanine + adenine equals 1. Since adenine equals thymine and guanine equals cytosine, it follows that the number of pyrimidines equals the number of purines. Each base is linked to a pentose sugar forming the backbone of each DNA strand.

191. The answer is D. *(Stryer, 4/e, pp 90–92.)* RNA, not DNA, is utilized by some viruses as genetic material. Retroviruses contain RNA in their virions, and eventually replicate through a DNA intermediate. Retroviruses, such as Rous sarcoma virus, can induce cancer. Poliovirus also transmits RNA as genetic material. Certain viruses such as ϕX174 virus use circular DNA that is single-stranded. DNA is found outside the chromosome in bacteria as plasmids, which have been adapted for use in recombinant DNA studies. Mitochondrial DNA is a source of extrachromosomal DNA in eukaryotic cells that codes for the synthesis of certain mitochondrial RNAs. DNA extracted from closely related species hybridizes extensively, which indicates widespread homology. As the evolutionary space between species increases, homology decreases.

192. The answer is A. *(Stryer, 4/e, pp 120–122, 792, 859.)* Restriction endonucleases are produced by prokaryotes for cleaving both strands of foreign DNA. The host cell's DNA is not degraded because the recognition sites are specifically methylated. The endonucleases recognize specific short symmetric sequences known as *palindromes.* These cleavage sites contain twofold rotational symmetry in that the sequence is identical but antiparallel in the complementary strands. In some cases, single-stranded, cohesive ends on each of the complementary strands are produced, while in other cases, double-stranded, blunt ends are formed. Modern analysis of DNA structure is highly dependent upon the use of different restriction endonucleases that permit the specific hydrolysis of DNA into large polynucleotides.

193. The answer is B. *(Stryer, 4/e, pp 122–124.)* DNA sequence analysis is often carried out using the specific chemical cleavage method worked out by

Maxam and Gilbert. In this method, restriction endonucleases are used to cleave DNA molecules into specific fragments. Each specific fragment is analyzed. The polynucleotide is labeled at one specific terminus (usually the 5′ end) with ^{32}P using polynucleotide kinase. Four different aliquots of the polynucleotide are preferentially cleaved at a different nucleotide (thymine, adenine, cytosine, and guanine). The reaction is timed to produce one break per chain. Depending upon where the cleavage has occurred, polynucleotides of different lengths will be produced and resolved by gel electrophoresis. The electrophoresis will separate molecules differing by just one nucleotide. Thus, a full spectrum of bands varying in length will be observed. The larger the band, the slower it will move in the gel. The bands can be identified by their radioactivity. By comparing the four gels, the sequence can be determined directly by the sequence in which the bands migrate.

194. The answer is C. *(Stryer, 4/e, pp 852–854, 903–906.)* RNA is synthesized from DNA as single-stranded, linear molecules. Since it is not double-stranded, the concentrations of the different bases in RNA are variable. The hybridization of RNA with its complementary template DNA is antiparallel. At physiologic pH, RNA (like DNA) is polyanionic owing to the negatively charged phosphate hydroxyl groups. In both DNA and RNA, sugar units equal base units equal phosphate units.

195. The answer is D. *(Stryer, 4/e, pp 894–901.)* The initiation but not the elongation of polypeptide chain synthesis (in *E. coli*) involves N-terminal *N*-formylmethionine incorporation as specified by codons AUG and GUG. Peptide-chain elongation on the ribosomal mRNA employs peptidyl transferase after codon-specific binding of aminoacyl-tRNA and uses energy from GTP and the factors G, Tu, and Ts.

196. The answer is A. *(Stryer, 4/e, pp 879–880, 894–901.)* Four high-energy phosphate bonds are utilized during the overall process of linking together two amino acids via formation of peptide bonds. The equivalents of two bonds from ATP are used during activation of amino acids, since ATP is hydrolyzed to AMP and inorganic pyrophosphate (PP_i). One bond from GTP is used during binding of aminoacyl-tRNA to the active ribosomal site, and another bond from GTP is hydrolyzed during translocation. During the process of initiation, an extra GTP is hydrolyzed during the attachment of *N*-formylmethionyl-tRNA to mRNA and the association of ribosomal subunits. The step in which actual formation of peptide bonds is catalyzed by peptidyl transferase requires no consumption of high-energy phosphate bonds.

197. The answer is C. *(Stryer, 4/e, pp 894–895, 903–905.)* Initiation of protein synthesis in *E. coli,* mitochondria, and chloroplasts requires *N*-formylmethionyl-tRNA association with ribosomes. In general, initiation of translation on cytoplasmic ribosomes of eukaryotes is similar. However, the special tRNA used in eukaryotes for initiation is methionyl-tRNA, which is not formylated.

198. The answer is A. *(Stryer, 4/e, pp 803, 806–809.)* DNA polymerase III catalyzes the step-by-step polymerization of deoxyribonucleoside 5′-triphosphate units (dATP, dTTP, dCTP, and dGTP) to the 3′-hydroxyl terminus of a preexisting RNA or DNA primer strand. During the course of this magnesium-requiring reaction, pyrophosphate is liberated. Pyrophosphatase quickly converts the pyrophosphate to inorganic phosphate.

$$(DNA)_n + dNTP \leftrightarrow (DNA)_{n+1} + PP_i$$

Since addition is at the free 3′-hydroxyl terminus, elongation of the new DNA chain is in the $5' \rightarrow 3'$ direction. DNA polymerase III prefers double-stranded DNA templates and is present as a multisubunit complex known as DNA polymerase III holoenzyme.

199. The answer is E. *(Stryer, 4/e, pp 128–137.)* The introduction of recombinant DNA into the host bacteria is usually accomplished by packaging it into infectious λ-bacteriophages. Another method is simply to allow the bacterial cells to take up naked recombinant DNA molecules (e.g., plasmids) from growth medium. The latter process occurs with a frequency of about one of every 10^6 DNA molecules. Conjugation is not used to introduce artificially produced recombinant DNA. Although a variety of methods exists, recombinant DNA may be formed by splitting plasmid or bacteriophage vehicle DNA as well as donor (e.g., eukaryotic DNA) with specific restriction endonucleases. The foreign donor and vehicle DNA is then annealed together with DNA ligase and introduced into the bacterial host cell. Marker genes, such as those for antibiotic resistance, may be included in the vehicle DNA so that the host cells containing the introduced recombinant DNA may be specifically selected and cloned.

200. The answer is D. *(Stryer, 4/e, pp 752–754.)* Aminopterin and methotrexate (amethopterin) are competitive inhibitors of dihydrofolate reductase. This action blocks the regeneration of tetrahydrofolate from dihydrofolate. Consequently, deoxythymidylate synthesis is interrupted. The action of the drugs derives from the fact that they are analogues of folic acid. Both drugs are used in the treatment of certain cancers such as leukemia. Likewise, fluorouracil is used in cancer chemotherapy and blocks the synthesis of deoxythymidylate. In vivo, fluorouracil is converted to fluorouridylate, which is a suicide inhibitor of thymidylate synthase. Thymidylate synthase catalyzes the reaction dUMP → dTMP. Chloramphenicol is an antibiotic inhibitor of protein synthesis that affects peptidyl transferase activity of prokaryotes as well as eukaryotes.

201. The answer is B. *(Stryer, 4/e, pp 58, 229, 384, 834.)* The AIDS treatment drug azidothymidine (AZT) exerts its effect by inhibiting reverse transcriptase. Thus, it prevents replication of the human immunodeficiency virus. Reverse transcriptase is an RNA-directed DNA polymerase. The RNA of retroviruses utilizes reverse transcriptase to synthesize DNA provirus, which, in turn, synthesizes new viral RNA. AZT does not directly inhibit synthesis of new viral RNA.

202. The answer is C. *(Stryer, 4/e, pp 104–110.)* Since there are 4 different bases that can specify amino acids as triplet codons, 64 codons are possible. Thus, most amino acids are specified by more than one codon, which makes the code degenerate. Only UAG, UAA, and UGA do not specify amino acids; they are termination signals, or stop codons. In the example used (UCCUUCUUC), the degeneracy of the code is seen, since UCC and UCU both specify Ser.

203. The answer is C. *(Stryer, 4/e, pp 126–128, 793–794.)* DNA ligase catalyzes the formation of a phosphodiester bond between two DNA chains that must be part of a double-stranded DNA molecule; the enzyme cannot link single-stranded DNA molecules. The linkage occurs between the 5′-phosphate at one molecule's end and the 3′-OH at the other chain's end. The energy source for the bond formation in *E. coli* is NAD^+; in some animal cells and bacteriophages, ATP serves this purpose. Ligase is essential for normal DNA synthesis, DNA repair, and DNA chain splicing in genetic recombination.

204. The answer is A. *(Stryer, 4/e, pp 755–757.)* Xanthine oxidase catalyzes the last two steps in the degradation of purines. Hypoxanthine is oxidized to xanthine, which is further oxidated to uric acid. Thus, xanthine is both product and substrate in this two-step reaction. In humans, uric acid is excreted via the urine. Allopurinol, an analogue of xanthine, is used in gout to block uric acid production and deposition of uric acid crystals in the kidney and joints. It acts as a suicide inhibitor of xanthine oxidase after it is converted to alloxanthine. Guanine can also be a precursor of xanthine.

205. The answer is E. *(Stryer, 4/e, pp 747–748.)* The steps of pyrimidine nucleotide biosynthesis are summarized below:

$$\begin{array}{r} \text{Glutamine} + CO_2 + \text{ATP}\text{—}1\rightarrow\text{Carbamoyl phosphate} \\ | \\ 2 \leftarrow \text{Aspartate} \\ | \\ \text{CTP} \leftarrow 5\text{—UTP} \leftarrow 4\text{—UMP} \leftarrow 3\text{—Carbamoyl aspartate} \end{array}$$

The first step in pyrimidine synthesis is the formation of carbamoyl phosphate. The enzyme of this step, carbamoyl phosphate synthetase (1), is feedback-inhibited by UMP. The enzyme of the second step, aspartate transcarbamoylase (2), is feedback-inhibited by CTP. ATP is an activator of this enzyme. Aspartate transcarbamoylase is composed of catalytic and regulatory subunits. The regulatory subunit binds CTP or ATP. TTP has no role in the feedback regulation of pyrimidine synthesis.

206. The answer is A. *(Stryer, 4/e, pp 104–112.)* Each group of three bases in a sequence codes for an amino acid. The next three bases specify another amino acid. Thus, the genetic code is nonoverlapping. The triplet genetic code is degenerate, which is to say that, for most amino acids, there is more than one code word. The triplets of bases (codons) that specify the same amino acid usually differ only in the last base of the triplet. Chain termination is determined by three codons: UAA, UAG, and UGA.

207. The answer is B. *(Stryer, 4/e, p 745.)* Several control sites exist in the path of purine synthesis where feedback inhibition occurs. AMP, GMP, or IMP may inhibit the first step of the pathway, which is the synthesis of 5-phosphoribosyl-1-pyrophosphate (PRPP). PRPP synthetase is specifically inhibited. All three nucleotides can inhibit glutamine PRPP aminotransferase, which catalyzes the second step of the pathway. AMP blocks the conversion of IMP to adenylosuccinate. GMP inhibits the formation of xanthylate from IMP. Thus, blockage of IMP metabolism to AMP and GMP effectively inhibits purine biosynthesis.

208. The answer is C. *(Stryer, 4/e, pp 893–902.)* Primer is not used in protein synthesis as it is in DNA synthesis. Prior to the start of protein synthesis, ATP is required for activation of amino acids. The activated aminoacyl-tRNAs then interact with ribosomes carrying mRNA. Peptidyl transferase catalyzes the formation of peptide bonds between the free amino group of activated aminoacyl-tRNA on the A site of the ribosome and the esterified carboxyl group of the peptidyl-rRNA on the P site; the liberated rRNA remains on the P site.

209. The answer is C. *(Stryer, 4/e, pp 756–658.)* The genetic deficiency that results in the absence of hypoxanthine-guanine phosphoribosyltransferase (HGPRT) is the Lesch-Nyhan syndrome, a disease that produces self-destructive behavior, mental retardation, and spasticity. In normal persons, the enzyme catalyzes the salvage synthesis of inosine 5′-monophosphate (IMP) or guanosine 5′-monophosphate (GMP) from the reaction of hypoxanthine or guanine with phosphoribosylpyrophosphate (PRPP). Although gout can be caused by a variety of less devastating biochemical lesions that lead to the precipitation of sodium urate crystals from elevated serum levels of urate, in the Lesch-Nyhan syndrome gout is but one of the many symptoms caused by elevated levels of urate and PRPP. In some forms of gout, as well as in the Lesch-Nyhan syndrome, treatment with allopurinol inhibits xanthine oxidase and thereby prevents overproduction of urate from excess hypoxanthine. However, the drug has no effect on increased PRPP levels, which accelerate de novo synthesis of purines. Allopurinol does not alleviate the devastating neurologic consequences of the syndrome.

210. The answer is D. *(Stryer, 4/e, pp 128–130, 826–827.)* Plasmids are small, circular accessory chromosomes that can replicate autonomously of the host chromosome or insert into the host chromosome. In contrast to general genetic recombination, where homologous sequences (alleles) are exchanged, plasmids containing insertion elements can join unrelated genes and thus effect recombination of nonhomologous DNA. Plasmids may also carry genes that confer maleness and allow conjugation (F factor), carry drug-resistance genes (R factors), or carry genes for toxin production (colicinogenic factors). Plasmids are one class of mobile genetic elements (transposons) that are normally found in bacteria. Experimental insertion of foreign DNA into plasmids allows cloning of recombinant DNA.

211. The answer is E. *(Stryer, 4/e, pp 886–888.)* Owing to "wobble," a substitution in the 3′ base of certain, but not all, codons will still be recognized by the same anticodon of tRNA. Thus, it is conceivable that a mutation that changes a base will have no effect on the primary structure of a protein. In other cases, a new and different amino acid may be substituted that is similar to the original amino acid or just happens to be in a noncritical region of the protein. In either of these cases, normal function will still occur. Finally, substitution or loss of an amino acid from a critical region will result in a protein with no function or altered function. However, in all cases some sort of peptide will be produced.

212. The answer is A. *(Stryer, 4/e, pp 75–77, 740.)* A nucleoside consists of a purine or pyrimidine base linked to a pentose sugar. In DNA, deoxyribose sugars are used; in RNA, ribose sugars are used. On the other hand, nucleotides are phosphate esters of nucleosides with one to three phosphate groups, such as ATP, ADP, or AMP. The nitrogenous bases are adenine, thymine, guanine, cytosine, or uridine.

213–214. The answers are 213-B, 214-C. *(Stryer, 4/e, pp 75–77.)* The structures of the following bases are shown: A = cytosine; B = uracil; C = thymine; D = adenine; E = guanine. The nitrogenous bases are aromatic compounds. The pyrimidines—uracil, thymine, and cytosine—contain one heterocyclic ring each. The purines—adenine and guanine—are derivatives of pyrimidines and consist of a pyrimidine ring joined with an imidazole ring. All the bases except for uracil are found in DNA, while uracil replaces thymine in RNA.

215–216. The answers are 215-C, 216-C. *(Stryer, 4/e, pp 951–957.)* Repressor is encoded by the regulator gene. The binding of repressor protein to the operator gene prevents transcription of the structural genes. When inducer molecules bind the repressor, it cannot interact with the operator and synthesis of RNA proceeds. Binding of repressor is thought to inhibit binding and function of RNA polymerase.

Inducible catabolic operons are regulated by cyclic AMP, which acts by binding CAP protein. The CAP protein then binds the promoter and stimulates the transcription of the structural genes. It is thought that the binding of CAP protein to operator creates additional binding sites for RNA polymerase and thereby enhances transcription. This may be contrasted to repressors, which interfere with binding of RNA polymerase.

217–220. The answers are 217-B, 218-D, 219-C, 220-A. *(Stryer, 4/e, pp 851–852, 902–903.)* Antibiotics are highly specific inhibitors of biochemical steps. Streptomycin and tetracycline inhibit translation. Both compounds bind ribosomes. Streptomycin binds the 30S subunit of ribosomes. In this way it inhibits initiation of mRNA synthesis. Streptomycin also causes misreading of codons in prokaryotes. Tetracycline also binds the 30S subunit of ribosomes. By binding in this position, it blocks subsequent binding of aminoacyl-tRNA to mRNA in prokaryotes.

In contrast, actinomycin D and rifamycin B are inhibitors of DNA-directed RNA synthesis. Actinomycin D is a polypeptide-containing antibiotic. Like rifamycin B it is produced by a strain of *Streptomyces.* Actinomycin prevents DNA from acting as a template for RNA synthesis. It binds the helix of DNA. Rifamycin blocks formation of the first phosphodiester bond of RNA and thus prevents the initiation of transcription. The β-subunit of RNA polymerase is the site of action of rifamycin B.

221–224. The answers are 221-A, 222-B, 223-A, 224-C. *(Stryer, 4/e, pp 740–749, 752–754.)* The rate-controlling step in the biosynthesis of pyrimidines is the formation of *N*-carbamoyl aspartate from aspartate and *N*-carbamoyl phosphate. The rate-limiting enzyme is aspartate transcarbamoylase (ATCase). ATCase is feedback-inhibited by CTP, which is the final product of the pathway. The pathway also produces UTP, the direct precursor of CTP. 5′-Phosphoribosyl-1-1-pyrophosphate (PRPP) donates the ribose phosphate unit of nucleotides and is absolutely required for the beginning of the synthesis of purines. In fact, the enzymes regulating the synthesis of PRPP and the subsequent synthesis of phosphoribosylamine from PRPP are all end-product inhibited by IMP, AMP, and GMP, the products of this reaction pathway. The degradation of purines to urate can lead to gout when an elevated level of urate is present in serum, causing the precipitation of sodium urate crystals in joints. The excessive production of urate in many patients seems to be connected to a partial deficiency of hypoxanthine-guanine phosphoribosyl transferase (HGPRT). Allopurinol, which is an analogue of hypoxanthine, is a drug used to correct gout. It does so by inhibiting the production of urate from hypoxanthine and in doing so undergoes suicide inhibition of xanthine oxidase. Xanthine oxidase catalyzes a two-step reaction that includes the production of xanthine from hypoxanthine and then urate from xanthine.

225–228. The answers are 225-C, 226-E, 227-G, 228-A. *(Stryer, 4/e, pp 121–122, 126–131, 136–137.)* In an expression library, cDNA clones are screened on the basis of their ability to direct bacterial synthesis of a foreign protein of interest. Radioactive antibodies specific to this protein can be used to identify the colonies of bacteria that contain the cDNA vector. As was the case for probing genomic libraries, bacteria grown on a master plate are blotted onto a nitrocellulose replica plate and then lysed. The released proteins may then be labeled with ^{125}I antibodies. In contrast, northern blotting can be used to identify RNA molecules separated by gel electrophoresis. In northern blotting, RNA molecules separated by gel electrophoresis can be identified by hybridization with probe DNA following transfer to nitrocellulose. Mutant and wild-type oligonucleotides can be used as probes to analyze polymerase chain reaction products. Conversely, the products of polymerase chain reaction can be used to analyze cDNA libraries.

229–230. The answers are 229-C, 230-B. *(Stryer, 4/e, pp 799–803, 982–983.)* The initiation of DNA synthesis in *E. coli* requires the unwinding of parental DNA and the formation of complementary RNA primers by DNA-directed RNA polymerase (primase). The actual replication of DNA is carried out by DNA polymerase III in bacteria and α-DNA polymerase in eukaryotic cells. In both prokaryotic cells and eukaryotic cells, DNA replication occurs at more than one site. Thus, the ends of the discontinuous pieces of DNA (Okazaki fragments) synthesized are joined together by DNA ligase to create a continuous, complete chain of DNA. Repair of DNA is carried out by β-DNA polymerase in eukaryotic cells.

Carbohydrates and Lipids

Note: P/O ratios used for ATP values are variable and in disagreement dependent upon the textbook used. Routinely, it has been taught that oxidation of glucose or other organic molecules yields 3 ATP per NADH, 2 ATP per $FADH_2$, and 12 ATP per acetyl CoA produced. For the sake of consistency, these established criteria are used in this review.

Regulatory Enzymes of Intermediary Metabolism

Enzyme	Allosteric Effectors*	Hormonal Effectors*
GLYCOLYSIS		
Hexokinase	Glucose-6-phosphate (−)	
Phosphofructokinase	ATP (−), citrate (−), fructose-2,6-bisphosphate (+)	Glucagon [liver (−)], epinephrine [liver (−) & muscle (+)]
Pyruvate kinase	ATP (−), alanine (−), fructose-1,6-bisphosphate (+)	Glucagon [liver (−)]
GLUCONEOGENESIS (LIVER)		
Pyruvate carboxylase	ADP (−), acetyl CoA (+)	
Phosphoenolpyruvate carboxykinase	ADP (−)	
Fructose-1,6-bisphosphatase	AMP (−), citrate (+), fructose-2, 6-bisphosphate (−)	Glucagon (+), epinephrine (+)
Glucose-6-phosphatase	Glucose (−)	
GLYCOGEN STORAGE AND BREAKDOWN		
Muscle glycogen phosphorylase	ATP (−), glucose-6-phosphate (−), AMP (+), Ca (+)	Epinephrine (+), insulin (−)
Liver glycogen phosphorylase	Glucose (−)	Glucagon (+), epinephrine (+), insulin (−)

*(−) negative effector; (+) positive effector

Regulatory Enzymes of Intermediary Metabolism

Enzyme	Allosteric Effectors*	Hormonal Effectors*
Glycogen synthase	Glucose (+)	Glucagon (−), epinephrine (−), insulin (+)
CITRIC ACID CYCLE		
Pyruvate dehydrogenase complex	ATP (−), acetyl CoA (−), NADH (−), insulin (+), AMP(+), CoA(+) NAD^+ (+), Ca (+)	
Citrate synthase	ATP (−)	
Isocitrate dehydrogenase	ATP (−), NADH (−), ADP (+)	
α-Ketoglutarate dehydrogenase	Succinyl CoA (−), NADH (−)	
OXIDATIVE PHOSPHORYLATION		
Electron transport and proton gradient	ADP (+)	
PENTOSE PHOSPHATE PATHWAY		
Glucose-6-phosphate dehydrogenase	NADPH (−), $NADP^+$ (+)	
FATTY ACID SYNTHESIS		
Acetyl CoA carboxylase	Palmitoyl CoA (−), citrate (+)	Glucagon (−), epinephrine (−), insulin (+)
FATTY ACID OXIDATION		
Carnitine acyltransferase I	Malonyl CoA (−)	
3-Hydroxyl CoA dehydrogenase	NADH (−)	
Thiolase	Acetyl CoA	
CHOLESTEROL SYNTHESIS		
3-Hydroxy-3-methylglutaryl CoA reductase	Cholesterol (−), mevalonate (−)	
LIPOLYSIS		
Hormone-sensitive triacylglycerol lipase		Insulin (−), glucagon (+), epinephrine (+)

DIRECTIONS: Each question below contains five suggested responses. Select the **one best** response to each question.

231. Chylomicra, intermediate-density lipoproteins (IDL), low-density lipoproteins (LDL), and very low-density lipoproteins (VLDL) all are serum lipoproteins. What is the correct ordering of these particles from the lowest to the greatest density?

(A) LDL, IDL, VLDL, chylomicra
(B) Chylomicra, VLDL, IDL, LDL
(C) VLDL, IDL, LDL, chylomicra
(D) Chylomicra, IDL, VLDL, LDL
(E) LDL, VLDL, IDL, chylomicra

232. The synthesis of fatty acids is often termed *reductive synthesis*. The reducing equivalent for fat production is provided by which of the following compounds?

(A) Nicotinamide adenine dinucleotide phosphate ($NADP^+$)
(B) Flavin adenine dinucleotide (FAD^+)
(C) The reduced form of flavin adenine dinucleotide ($FADH_2$)
(D) The reduced form of nicotinamide adenine dinucleotide phosphate (NADPH)
(E) The reduced form of nicotinamide adenine dinucleotide (NADH)

233. Given that the standard free energy change ($\Delta G°'$) for the hydrolysis of ATP is −7.3 kcal/mol and that for the hydrolysis of glucose-6-phosphate is −3.3 kcal/mol, what is the $\Delta G°'$ for the phosphorylation of glucose?

Glucose + ATP →
Glucose-6-phosphate + ADP

(A) −10.6 kcal/mol
(B) −7.3 kcal/mol
(C) −4.0 kcal/mol
(D) +4.0 kcal/mol
(E) +10.6 kcal/mol

234. A low level of carbon dioxide labeled with ^{14}C is accidentally released into the atmosphere surrounding industrial workers as they resume work following the lunch hour. Unknowingly, they breathe the contaminated air for 2 h. Which of the following compounds is labeled?

(A) Acetyl CoA
(B) About one-third the carbon atoms of malonyl CoA
(C) About one-half the carbon atoms of some fatty acids
(D) The carboxyl atom of fatty acids
(E) All the carbon atoms of fatty acids

235. What is the structure shown?

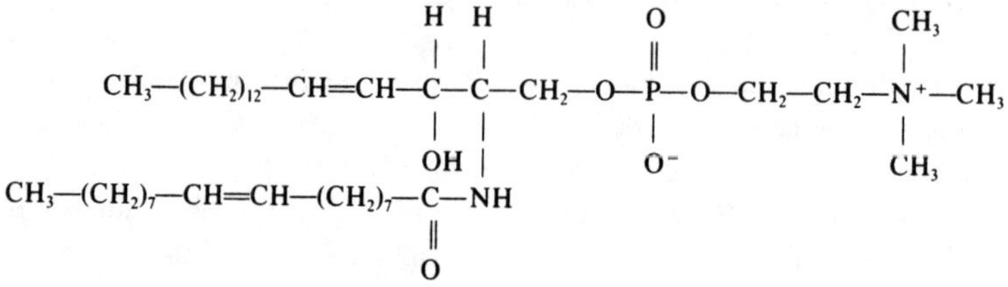

```
                                    H   H                  O                       CH3
                                    |   |                  ||                      |
          CH3—(CH2)12—CH=CH—C—C—CH2—O—P—O—CH2—CH2—N+—CH3
                                    |   |                  |                       |
                                    OH  |                  O−                      CH3
CH3—(CH2)7—CH=CH—(CH2)7—C—NH
                                    ||
                                    O
```

(A) Cerebroside
(B) Sphingomyelin
(C) Ganglioside
(D) Phosphatidyl choline
(E) Phosphatidyl ethanolamine

---

236. The key regulatory enzyme of the pentose phosphate pathway is positively regulated by

(A) NADH
(B) ADP
(C) GTP
(D) $NADP^+$
(E) FADH

237. In the presence of rotenone,

(A) NADH is oxidized by electron transport
(B) cytochrome *a* is reduced by electron transport
(C) cytochrome *b* is reduced by electron transport
(D) cytochrome *c* is reduced by electron transport
(E) $FADH_2$ is oxidized by electron transport

238. Which of the following occurs in nonshivering thermogenesis?

(A) Glucose is oxidized to lactate
(B) Fatty acids uncouple oxidative phosphorylation
(C) Ethanol is formed
(D) ATP is burned for heat production
(E) Adipose tissue is functionally absent

239. Aspirin inhibits which of the following enzymes?

(A) Lipoprotein lipase
(B) Lipoxygenase
(C) Cyclooxygenase
(D) Phospholipase D
(E) Phospholipase A_2

240. Which of the following steps in the biosynthesis of cholesterol is thought to be rate-controlling and the locus of metabolic regulation?

(A) Geranyl pyrophosphate → farnesyl pyrophosphate
(B) Squalene → lanosterol
(C) Lanosterol → cholesterol
(D) 3-Hydroxy-3-methylglutaryl CoA → mevalonic acid
(E) None of the above

241. The correct order of passage of electrons through the cytochromes of the respiratory chain is

(A) $a \to a_3 \to b \to c \to c_1$
(B) $a_3 \to a \to b \to c_1 \to c$
(C) $b \to c_1 \to c \to a \to a_3$
(D) $a_3 \to a \to c \to c_1 \to b$
(E) $c \to b \to c_1 \to a_3 \to a$

242. The structure shown below is that of

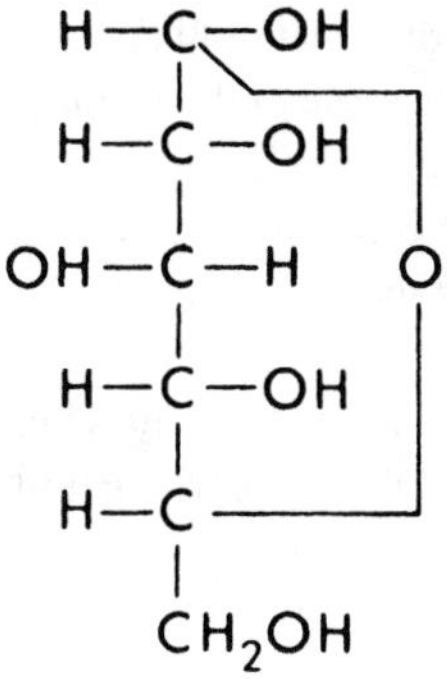

```
    H—C—OH ─────┐
      |         |
    H—C—OH      |
      |         |
   OH—C—H       O
      |         |
    H—C—OH      |
      |         |
    H—C ────────┘
      |
     CH2OH
```

(A) α-D-glucopyranose
(B) β-D-glucopyranose
(C) α-D-glucofuranose
(D) β-L-glucofuranose
(E) α-D-fructofuranose

243. Chronic alcoholics require more ethanol than do nondrinkers to become intoxicated because of a higher level of specific enzyme. However, independent of specific enzyme levels, the availability of what other substance is rate-limiting in the clearance of ethanol?

(A) NADH
(B) NAD^+
(C) FADH
(D) FAD^+
(E) NADPH

244. A choline residue is present in which of the following lipids?

(A) Phosphatidic acid
(B) Ganglioside
(C) Cholesterol
(D) Glucocerebroside
(E) Sphingomyelin

245. Which of the following processes will yield arachidonic (5,8,11,14-eicosatetraenoic) acid in mammals?

(A) Elongation of stearic acid
(B) Chain elongation and one desaturation of linolenic (9,12,15-octadecatrienoic) acid
(C) Chain elongation and two desaturations of linoleic (9,12-octadecadienoic) acid
(D) Desaturation of oleic acid
(E) Elongation of palmitic acid

246. All known effects of cyclic AMP in eukaryotic cells result from

(A) activation of the catalytic unit of adenylate cyclase
(B) activation of synthetases
(C) activation of protein kinase
(D) phosphorylation of G protein
(E) stimulation of Ca^{2+} release from endoplasmic reticulum

247. Cyanides produce hypoxia by

(A) producing central hypoventilation
(B) interfering with oxygen carriage
(C) slowing capillary circulation
(D) inhibiting cellular respiration
(E) none of the above mechanisms

248. The Mitchell or chemiosmotic hypothesis predicts the

(A) existence of a higher pH in the cisternae of the endoplasmic reticulum than in the cytosol
(B) synthesis of ATP as protons flow into the mitochondrial matrix along a proton gradient that exists across the inner mitochondrial membrane
(C) symmetric distribution of the ATPase of the inner membrane of the mitochondria
(D) dissociation of electron transport and oxidative phosphorylation
(E) absence of ATPase in the inner membrane of the mitochondria

249. How many moles of high-energy phosphate bond equivalents are utilized in the synthesis of 1 mole of triacylglycerol from free fatty acids and glycerol?

(A) 3
(B) 4
(C) 7
(D) 9
(E) 12

250. The Pasteur effect refers to

(A) an increase in hexokinase activity owing to increased oxidative phosphorylation
(B) an increase in ethanol formation from pyruvate on changing from anaerobic to aerobic metabolism
(C) an increase in glucose utilization via the pentose phosphate pathway on changing from anaerobic to aerobic metabolism
(D) a decrease in glucose utilization on changing from anaerobic to aerobic metabolism
(E) a decrease in the respiratory quotient on changing from carbohydrate to fat as the major metabolic fuel

251. Which of the following enzymes of the glycolytic pathway is particularly sensitive to inhibition by fluoride ions?

(A) Hexokinase
(B) Aldolase
(C) Pyruvate kinase
(D) Enolase
(E) Phosphohexose isomerase

252. If all potential sources of ATP production are taken into account, the net number of ATP molecules formed per molecule of glucose in aerobic glycolysis is

(A) 2
(B) 6
(C) 18
(D) 36
(E) 54

253. Among the many molecules of high-energy phosphate compounds formed as a result of the functioning of the citric acid cycle, one molecule is synthesized at the substrate level. In which of the following reactions does this occur?

(A) Citrate → α-ketoglutarate
(B) α-Ketoglutarate → succinate
(C) Succinate → fumarate
(D) Fumarate → malate
(E) Malate → oxaloacetate

254. The fatty acid synthase complex of mammals

(A) is a dimer of unsimilar subunits
(B) is composed of seven different proteins
(C) dissociates into eight different proteins
(D) catalyzes eight different enzymatic steps
(E) is composed of covalently linked enzymes

255. The process of glycolysis includes the following reactions and their concomitant free-energy changes:

Glyceraldehyde-3-phosphate + $NAD^+ + P_i \rightleftharpoons$ 1,3-diphosphoglycerate + NADH + H^+: $\Delta G^{\circ\prime} = +1.5$ kcal/mol

1,3-Diphosphoglycerate + ADP $\rightleftharpoons$ 3-phosphoglycerate + ATP: $\Delta G^{\circ\prime} = -4.5$ kcal/mol

For the two-step process converting glyceraldehyde-3-phosphate to 3-phosphoglycerate, the overall free-energy change is

(A) $\Delta G^{\circ\prime} = +6.0$ kcal/mol
(B) $\Delta G^{\circ\prime} = +3.0$ kcal/mol
(C) $\Delta G^{\circ\prime} = -3.0$ kcal/mol
(D) $\Delta G^{\circ\prime} = -4.5$ kcal/mol
(E) $\Delta G^{\circ\prime} = -6.0$ kcal/mol

256. The activity of pyruvate carboxylase is dependent upon the positive allosteric effector

(A) succinate
(B) AMP
(C) isocitrate
(D) citrate
(E) acetyl CoA

257. In the figure shown below, fructose-1,6-diphosphate is located at point

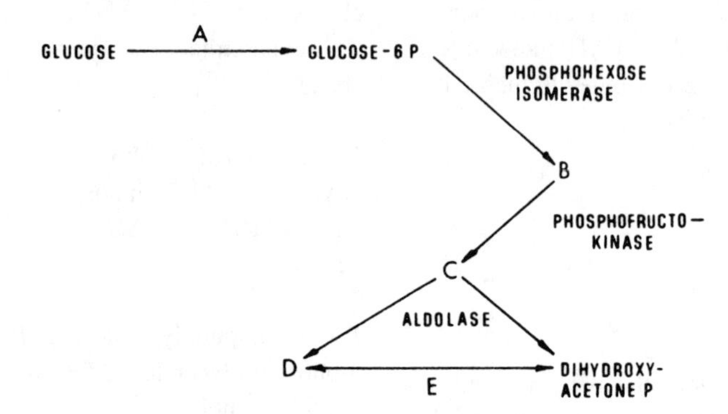

(A) A
(B) B
(C) C
(D) D
(E) E

---

258. Which of the following statements best describes the function of carnitine?

(A) It transports medium-chain fatty acids into gut epithelial cells
(B) It transports medium-chain fatty acids across the inner mitochondrial membrane
(C) It is a derivative of vitamin A and is involved in adaptation of the retina to darkness
(D) It is a transacylation reaction catalyzed by a transferase
(E) It is a coenzyme needed for one of the enzymatic steps in the synthesis of fatty acids

259. Following a diet fad meal of skim milk and yogurt, an adult female patient experiences abdominal distention, nausea, cramping, and pain followed by a watery diarrhea. This set of symptoms has been observed each time the meal is consumed. A likely diagnosis is

(A) steatorrhea
(B) lactase deficiency
(C) maltose deficiency
(D) sialilase deficiency
(E) lipoprotein lipase deficiency

260. The primary biochemical lesion in homozygotes with familial hypercholesterolemia (type IIa) is

(A) the loss of feedback inhibition of liver hydroxymethylglutaryl CoA reductase
(B) the increased production of low-density lipoproteins from very low-density lipoproteins
(C) the loss of apolipoprotein B
(D) the malfunctioning of acyl CoA–cholesterol acyl transferase (ACAT)
(E) the functional deficiency of plasma membrane receptors for low-density lipoproteins

261. In the reaction below, NuDP stands for

NuDP glucose + glycogen$_n$ → NuDP + glycogen$_{n+1}$

(A) ADP
(B) CDP
(C) GDP
(D) TDP
(E) UDP

262. In the reaction below, NuTP stands for

NuTP + glucose → glucose-6-phosphate + NuDP

(A) ATP
(B) CTP
(C) GTP
(D) TTP
(E) UTP

263. Rats fed a fat-free diet from birth would be deficient in

(A) sphingolipids
(B) phospholipids
(C) triacylglycerols
(D) prostaglandins
(E) cholesterol

264. In the pathway leading to biosynthesis of acetoacetate from acetyl CoA in liver, the immediate precursor of acetoacetate is which of the following substances?

(A) 3-Hydroxybutyrate
(B) Acetoacetyl CoA
(C) 3-Hydroxybutyryl CoA
(D) Mevalonic acid
(E) 3-Hydroxy-3-methylglutaryl CoA

265. Which of the following intermediates can be isolated from yeast that is fermenting wine but not from healthy muscle?

(A) Acetyl CoA
(B) Lactate
(C) Acetaldehyde
(D) Citrate
(E) Oxaloacetate

266. Which of the following statements correctly describes cytochrome oxidase?

(A) It is composed of one cytochrome
(B) It transfers four electrons and four protons to produce water
(C) It transfers electrons from cytochrome Q to Q_2
(D) It is inhibited by copper
(E) It is also known as *cytochrome c*

267. A compound normally used to conjugate bile acids is

(A) serine
(B) glucuronic acid
(C) fatty acids
(D) calcium
(E) glycine

268. In the pathway below, ATP is produced between

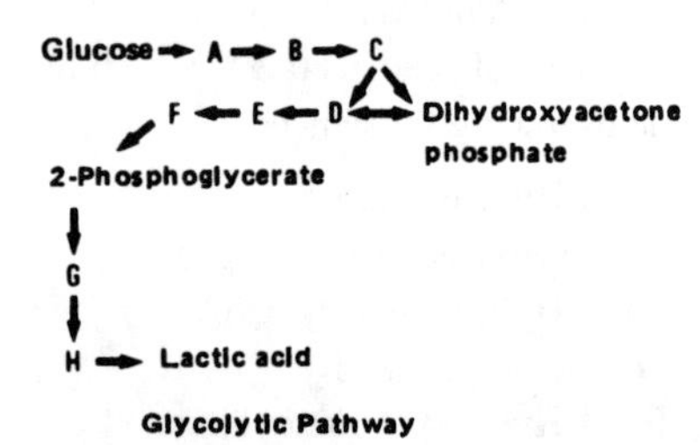

(A) A-B
(B) B-C
(C) C-D
(D) D-E
(E) G-H

269. Fatty acids that must be obtained from the diet by humans include

(A) oleic
(B) palmitoleic
(C) palmitic
(D) stearic
(E) linoleic

270. Under normal metabolic conditions, erythrocytes would be expected to accumulate

(A) phosphoenolpyruvate
(B) lactate
(C) NADPH
(D) citrate
(E) ethanol

271. The liver converts ethanol to

(A) methanol
(B) acetone
(C) acetaldehyde
(D) hydrogen peroxide
(E) glycerol

272. Which one of the following apolipoproteins is synthesized in liver as part of the coat of very low-density lipoproteins (VLDLs)?

(A) A-I
(B) B-48
(C) C-II
(D) B-100
(E) E

273. Which one of the following enzymes catalyzes high-energy phosphorylation of substrates during glycolysis?

(A) Pyruvate kinase
(B) Phosphoglycerate kinase
(C) Triose phosphate isomerase
(D) Aldolase
(E) Glyceraldehyde-3-phosphate dehydrogenase

274. Which of the following occurs to fructose during catabolism?

(A) It is phosphorylated in the first step
(B) After initial modification, it is cleaved by a specific enolase
(C) It is converted to UDP-fructose
(D) It is ultimately converted to galactose
(E) It is a substrate for fructose-6-phosphatase

275. Which one of the following reactions constitutes an isomerization?

(A) 1,3-Biphosphoglycerate ↔ 3-phosphoglycerate
(B) 2-Phosphoglycerate ↔ 3-phosphoglycerate
(C) 2-Phosphoglycerate ↔ phosphoenolpyruvate + H_2O
(D) Glucose-6-phosphate ↔ fructose-6-phosphate
(E) Glucose ↔ glucose-6-phosphate

276. Which of the following lipoproteins would contribute to a measurement of plasma cholesterol in a normal person following a 12-h fast?

(A) Very low-density lipoproteins
(B) High-density lipoproteins
(C) Chylomicra
(D) Chylomicron remnants
(E) Adipocyte lipid droplets

277. Which one of the following products of triacylglycerol breakdown and subsequent β-oxidation may undergo gluconeogenesis?

(A) Propionyl CoA
(B) Acetyl CoA
(C) All ketone bodies
(D) Some amino acids
(E) β-Hydroxybutyrate

278. Citrate has a positive allosteric effect on which one of the following enzymes?

(A) Pyruvate kinase
(B) Acetyl CoA carboxylase
(C) Phosphofructokinase
(D) Fatty acid synthetase
(E) Enolase

279. Which one of the following enzymes is common to both glycolysis and gluconeogenesis?

(A) Pyruvate kinase
(B) Pyruvate carboxylase
(C) Hexokinase
(D) Phosphoglycerate kinase
(E) Fructose-1,6-bisphosphatase

280. Which of the following regulates lipolysis in adipocytes?

(A) Activation of fatty acid synthesis mediated by cyclic AMP
(B) Activation of triglyceride lipase as a result of hormone-stimulated increases in cyclic AMP levels
(C) Glycerol phosphorylation to prevent futile esterification of fatty acids
(D) Activation of cyclic AMP production by insulin
(E) Hormone-sensitive lipoprotein lipase

281. Which of the following statements correctly describes the enzyme thiokinase?

(A) It yields acetyl CoA as a product
(B) It yields ADP as a product
(C) It yields CoA as a product
(D) It forms CoA thioesters as a product
(E) It requires β-ketoacyl CoA as a substrate

282. Inhibition of the synthesis of ATP during oxidative phosphorylation by oligomycin is thought to be due to

(A) blocking the proton gradient between NADH-Q reductase and QH_2
(B) blocking the proton gradient between cytochrome c_1 and cytochrome *c*
(C) dissociating cytochrome *c* from mitochondrial membranes
(D) inhibiting mitochondrial ATPase (ATP synthase)
(E) uncoupling electron transfer between NADH and flavoprotein

283. In terms of high-energy molecules, what is the cost of retrieving one end-glucose residue attached to glycogen by α-1,4 linkage?

(A) 1 ATP
(B) 2 ATP
(C) 1 NADH
(D) 1 NADPH
(E) No relative cost

284. Leukocyte samples isolated from the blood of a newborn infant are homogenized and incubated with ganglioside G_{M2}. Approximately 47 percent of the expected normal amount of *N*-acetylgalactosamine is liberated during the incubation period. These results indicate that the infant

(A) is a genetic carrier for Tay-Sachs disease
(B) is homozygous for the Tay-Sachs trait
(C) has Tay-Sachs syndrome
(D) will most likely be mentally retarded
(E) has relatively normal β-*N*-acetylhexosaminidase activity

285. Reduction of which one of the following substrates leads to a reducing equivalent in a step of the citric acid cycle?

(A) Succinyl CoA
(B) Malate
(C) Fumarate
(D) Oxaloacetate
(E) Citrate

286. Glycogen synthetase, the enzyme involved in the biosynthesis of glycogen, may

(A) be activated by the phosphorylation of a specific serine residue
(B) be activated by increased calcium levels
(C) be more specifically defined as *UDP-glucose-glycogen glucosyl transferase*
(D) synthesize glycogen without a polymer primer
(E) employ UDP-D-glucose as a glucosyl donor in both plants and animals

287. During the first week of a diet of 1500 calories per day, the oxidation of glucose via glycolysis in the liver of a normal 59-kg (130-lb) woman is inhibited by the lowering of which one of the following?

(A) Citrate
(B) ATP
(C) Fatty acyl CoA
(D) Ketone bodies
(E) Fructose-2,6-bisphosphate

288. Most of the reducing equivalents utilized for synthesis of fatty acids can be generated from

(A) the pentose phosphate pathway
(B) glycolysis
(C) the citric acid cycle
(D) mitochondrial malate dehydrogenase
(E) citrate lyase

289. The entry point into the citric acid cycle for isoleucine, valine, and the product of odd-chain fatty acids is

(A) fumarate
(B) pyruvate
(C) oxaloacetate
(D) citrate
(E) succinyl CoA

290. If all the enzymes, intermediates, and cofactors of the citric acid cycle as well as an excess of the starting substrate acetyl CoA were present and functional in an organelle-free solution at the appropriate pH, which of the following factors of the citric acid cycle would prove to be rate-limiting?

(A) Reduction of cofactors
(B) Half-life of enzymes
(C) Molecular oxygen
(D) Turnover of intermediates
(E) CoA

291. Which one of the following activities is simultaneously stimulated by epinephrine in muscle and inhibited by epinephrine in liver?

(A) Fatty acid oxidation
(B) Glycogenolysis
(C) Cyclic AMP synthesis
(D) Glycolysis
(E) Activation of phosphorylase

292. Lack of the process known as *α-oxidation* leads to

(A) accumulation of phytanic acid
(B) oxidation of branched-chain fatty acids
(C) formation of dicarboxylic acid
(D) formation of propionic acid
(E) cure of Refsum's disease

293. Synthesis of triacylglycerides in mammals

(A) requires acyl-carrier protein
(B) uses glycerol phosphate as a precursor
(C) can utilize CDP-diacylglycerol as a precursor
(D) is increased in the presence of high levels of cyclic AMP
(E) esterifies only 20 carbon fatty acids

294. Familial fructokinase deficiency causes no symptoms because

(A) hexokinase can phosphorylate fructose
(B) most tissues utilize fructose
(C) liver fructose-1-P aldolase is still active
(D) excess fructose does not escape into the urine
(E) excess fructose spills into the bowel and is eliminated in feces

295. Familial galactosemia is toxic because

(A) excess glucose accumulates in the blood
(B) galactose is converted to the toxic substance galactitol (dulcitol)
(C) elimination of milk from the diet is ineffective
(D) galactose is itself toxic in even small amounts
(E) glucose metabolism is shut down by excess galactose

296. For every 2 moles of free glycerol released by lipolysis of triacylglycerides in adipose tissue,

(A) 2 moles of triacylglycerides are released
(B) 2 moles of free fatty acids are released
(C) 1 mole of glucose can be synthesized in gluconeogenesis
(D) 1 mole of triacylglyceride is released
(E) 3 moles of acyl CoA are produced

297. The yield from complete oxidation of glycogen is approximately 4 kcal/g. However, under physiologic conditions, glycogen is highly hydrated, such that the true physiologic yield is only approximately 1.5 kcal/g. Under similar physiologic conditions, what is the approximate yield from the oxidation of triacylglyceride stores?

(A) 1 kcal/g
(B) 2 kcal/g
(C) 4 kcal/g
(D) 9 kcal/g
(E) 24 kcal/g

298. The structure in the diagram below is most completely and correctly described as a

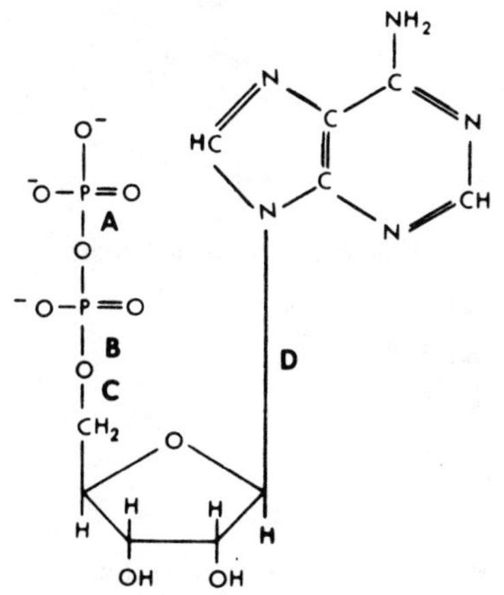

(A) base
(B) nucleoside
(C) nucleotide
(D) ribonucleotide
(E) deoxyribonucleotide

299. Which one of the following compounds is common to both the oxidative branch and the nonoxidative branch of the pentose phosphate pathway?

(A) Xylulose-5-phosphate
(B) Glucose-6-phosphate
(C) Glyceraldehyde-3-phosphate
(D) Fructose-6-phosphate
(E) Ribulose-5-phosphate

300. The reactions leading to the synthesis of squalene (C_{30}) from dimethylallyl pyrophosphate (C_5) are

(A) sequential condensation of 5-carbon units
(B) sequential condensation then cyclization of 5-carbon units
(C) sequential condensation of 5-carbon-pyrophosphate units
(D) sequential condensation of 5-carbon units to 15-carbon units, then condensation of 15-carbon units
(E) sequential condensation of 5-carbon units to 10-carbon units, then sequential condensation of 10-carbon units

301. A Nigerian medical student studying in the United States develops hemolytic anemia after taking the oxidizing antimalarial drug pamaquine. This severe reaction is most likely due to

(A) glucose-6-phosphate-dehydrogenase deficiency
(B) concomitant scurvy
(C) vitamin C deficiency
(D) diabetes
(E) glycogen phosphorylase deficiency

302. Transfer of H^+/e^- pairs to electron transport carriers, decarboxylation, and substrate-level phosphorylation occur at some of the steps shown in the following diagram of the citric acid cycle. All three of these events occur at step

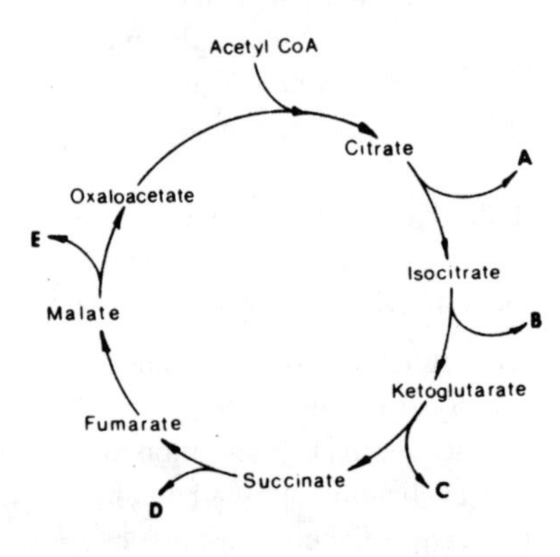

(A) A
(B) B
(C) C
(D) D
(E) E

303. In humans, the formation of the fatty acid C-18-Δ^9,Δ^{12} can be derived from which of the following?

(A) C-18 *cis*-Δ^9
(B) C-18 *cis*-Δ^6
(C) C-18
(D) C-16 *cis*-Δ^6,Δ^9
(E) None of the above

304. Which reaction in the figure shown below occurs in both muscle and liver but has substantially different qualities in the two?

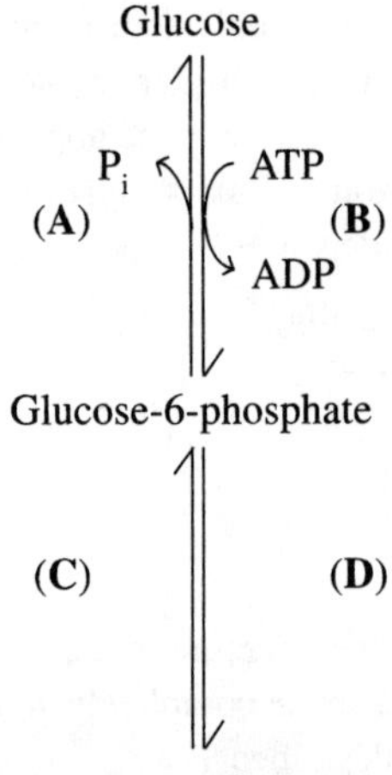

Fructose-6-phosphate

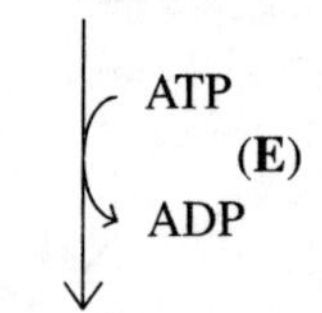

Fructose-1,6-bisphosphate

(A) A
(B) B
(C) C
(D) D
(E) E

305. Which one of the following enzymes catalyzes phosphorylation with the use of inorganic phosphate?

(A) Hexokinase
(B) Phosphofructokinase
(C) Glyceraldehyde-3-phosphate dehydrogenase
(D) Phosphoglycerate kinase
(E) Pyruvate kinase

306. Nervous stimulation of skeletal muscle causes the release of calcium from sarcoplasmic reticulum and leads to muscle contraction. Simultaneously, the increased calcium concentration causes

(A) a dramatic rise in cyclic AMP levels
(B) inactivation of glycogen phosphorylase
(C) activation of phosphorylase kinase
(D) activation of cyclic AMP phosphodiesterase
(E) activation of protein phosphatase

307. Which one of the following compounds is a key intermediate in the synthesis of both triacylglycerols and phospholipids?

(A) CDP-choline
(B) Phosphatidate
(C) Triacylglyceride
(D) Phosphatidylserine
(E) CDP-diacylglycerol

DIRECTIONS: Each numbered question or incomplete statement below is NEGATIVELY phrased. Select the **one best** lettered response.

308. All the following are necessary for the synthesis of fatty acids EXCEPT

(A) acetyl CoA
(B) NADPH + H^+
(C) ATP
(D) $FADH_2$
(E) HCO_3^-

309. All the following compounds can be considered potential precursors of gangliosides EXCEPT

(A) palmitoyl CoA
(B) cerebroside
(C) serine
(D) sphingomyelin
(E) sphingosine

310. During the process of glycoprotein formation, oligosaccharide units can be attached to the oxygen groups of all the following molecules EXCEPT

(A) dolichol phosphate
(B) tyrosine
(C) serine
(D) threonine
(E) asparagine

311. The enzymes α-dextrinase, α-amylase, and maltase are secreted by the salivary gland and the pancreas to hydrolyze dietary carbohydrates to yield all the following compounds EXCEPT

(A) fructose
(B) glucose
(C) dextrin
(D) maltose
(E) maltotriose

312. All the following effects of the steroid digitalis are observed following treatment of the survivors of congestive heart failure EXCEPT

(A) increase in cytosolic sodium levels
(B) inhibition of Na^+,K^+-ATPase
(C) increase in force of heart muscle contraction
(D) inhibition of plasma membrane ion pump
(E) decrease in cytosolic calcium

313. All the following are composed exclusively of glucose EXCEPT

(A) amylose
(B) lactose
(C) cellulose
(D) maltose
(E) glycogen

314. All the following statements about glycogen metabolism are true EXCEPT that

(A) cyclic AMP–activated protein kinase stimulates glycogen synthase
(B) phosphorylase kinase is activated by phosphorylation
(C) phosphorylase *b* is activated by phosphorylation
(D) cyclic AMP levels are raised by epinephrine and glucagon stimulation of adenylate cyclase
(E) a futile cycle of glycogenesis and glycogenolysis is prevented by second-messenger regulation

315. After a well-rounded breakfast, all the following would be expected to occur EXCEPT

(A) decreased activity of pyruvate carboxylase
(B) increased activity of acetyl CoA carboxylase
(C) increased rate of glycogenolysis
(D) increased rate of protein synthesis
(E) decreased activity of phosphoenolpyruvate carboxykinase

316. All the following are common to the synthesis of all steroid hormones EXCEPT

(A) cortisol
(B) progesterone
(C) cholesterol
(D) pregnenolone
(E) NADPH

317. All the following are constituents of ganglioside molecules EXCEPT

(A) glycerol
(B) sialic acid
(C) hexose sugar
(D) sphingosine
(E) long-chain fatty acid

318. All the following can serve as precursors for some prostaglandins and thromboxanes EXCEPT

(A) γ-linolenic acid
(B) linoleic acid
(C) eicosatrienoic acid
(D) eicosanoic acid
(E) eicosatetraenoic acid

319. All the following are components of prostaglandins EXCEPT

(A) hydroxyl and oxygen groups
(B) carboxyl group
(C) double bond
(D) methyl branch
(E) carbon ring

320. Sphingosine is the backbone of all the following EXCEPT

(A) lecithin
(B) ganglioside
(C) ceramide
(D) cerebroside
(E) sphingomyelin

321. All the following involve isoprenoids EXCEPT

(A) the chromophore of visual pigments
(B) cholesterol
(C) β-carotene
(D) the color of tomatoes
(E) ketone bodies

322. Positive signals for glycogen breakdown include increases in all the following EXCEPT

(A) cyclic AMP
(B) phosphorylated phosphorylase kinase
(C) epinephrine
(D) Ca^{2+}
(E) blood glucose

323. All the following statements about mammalian energy metabolism are true EXCEPT

(A) ATP is formed in the absence of Q_2
(B) ATP hydrolysis is an exergonic reaction
(C) ATP is formed in the presence of Q_2
(D) heat produced by ATP hydrolysis specifically drives other reactions
(E) NADH can be utilized to form ATP

324. During each cycle of ongoing β-fatty acid oxidation, all the following compounds are generated EXCEPT

(A) H_2O
(B) acetyl CoA
(C) fatty acyl CoA
(D) NADH
(E) $FADH_2$

325. Which one of the following reactions is LEAST likely to occur in cells?

(A) $AMP + PP_i \rightarrow ATP$
(B) $AMP + ATP \rightarrow 2\ ADP$
(C) $CDP + ATP \rightarrow CTP + ADP$
(D) $ADP + P_i \rightarrow ATP$
(E) $GDP + ATP \rightarrow GTP + ADP$

326. All the following statements about flavoproteins are true EXCEPT

(A) some are oxidized by coenzyme Q
(B) they receive electrons from cytochrome P_{450} in liver mitochondria
(C) they may be NADH dehydrogenases
(D) they can be associated with sulfur and nonheme iron
(E) they can produce hydrogen peroxide

327. All the following events occur during formation of phosphoenolpyruvate from pyruvate during gluconeogenesis EXCEPT

(A) CO_2 is consumed
(B) CO_2 is given off
(C) CoA is required
(D) ATP is hydrolyzed
(E) GTP is hydrolyzed

328. All the following statements describing lipids are true EXCEPT

(A) they usually associate by covalent interactions
(B) they are structural components of membranes
(C) they may contain nitrogen and phosphate in addition to carbon, hydrogen, and oxygen
(D) they are an intracellular energy source
(E) they are poorly soluble in water

329. All the following statements correctly describe ketone bodies EXCEPT

(A) they may result from starvation
(B) they are present at high levels in uncontrolled diabetes mellitus
(C) they are utilized by the liver during long-term starvation
(D) they include β-hydroxybutyrate and acetone
(E) they may be excreted in the urine

330. All the following are important intermediates in the synthesis of fatty acids from glucose EXCEPT

(A) mitochondrial acetyl CoA
(B) citrate
(C) cytosolic acetyl CoA
(D) 3-ketoacyl CoA
(E) malonyl CoA

331. All the following are needed to store blood glucose as glycogen following a meal EXCEPT

(A) glycogen-branching enzyme
(B) UTP
(C) glucose-6-phosphatase
(D) glucose-6-phosphate
(E) glucose-1-phosphate

332. Hydrolysis of a mixture of phosphoglycerides may yield all the following EXCEPT

(A) serine
(B) choline
(C) galactose
(D) glycerol
(E) phosphate

333. All the following correctly describe the intermediate 3-hydroxy-3-methylglutaryl CoA EXCEPT

(A) it is generated enzymatically in the mitochondrial matrix
(B) it is formed in the cytoplasm
(C) it inhibits the first step in cholesterol synthesis
(D) it is involved in the synthesis of ketone bodies
(E) it is an intermediate in the synthesis of cholesterol

334. Glucose-6-phosphate is a substrate or product in all the following pathways EXCEPT

(A) gluconeogenesis
(B) glycolysis
(C) glycogenolysis
(D) pentose phosphate shunt
(E) oxidative phosphorylation

335. Intermediates in the de novo pathway of synthesis of triacylglycerols include all the following EXCEPT

(A) fatty acyl CoA
(B) CDP-diacylglycerol
(C) glycerol-3-phosphate
(D) lysophosphatidic acid
(E) phosphatidic acid

336. Components of bile that are active in the intestinal digestive process include all the following EXCEPT

(A) taurocholate
(B) glycocholate
(C) deoxycholic acid
(D) bile pigments
(E) lecithin

337. The citric acid cycle is inhibited by all the following EXCEPT

(A) fluoroacetate
(B) fluorouracil
(C) anaerobic conditions
(D) arsenite
(E) malonate

338. All the following statements about the structure of glycogen are true EXCEPT

(A) it is a copolymer of glucose and galactose
(B) branched chains occur about every 10 residues
(C) it contains α-1,4 glycosidic linkages
(D) it contains α-1,6 glycosidic linkages
(E) monosaccharide residues are all D-glucose

339. Mitochondrial α-ketoglutarate dehydrogenase complex requires all the following to function EXCEPT

(A) CoA
(B) FAD
(C) NAD^+
(D) $NADP^+$
(E) thiamine pyrophosphate

340. Each of the following can be an intermediate in the synthesis of phosphatidyl choline EXCEPT

(A) CDP-choline
(B) phosphatidyl inositol
(C) phosphatidyl ethanolamine
(D) diacylglycerol
(E) phosphorylcholine

341. All the following contain glucose as a structural unit EXCEPT

(A) maltose
(B) glycogen
(C) mannose
(D) lactose
(E) sucrose

342. All the following are allosteric effectors regulating key irreversible enzymes of the glycolytic pathway EXCEPT

(A) AMP
(B) ADP
(C) ATP
(D) glucose-6-phosphate
(E) glucose

DIRECTIONS: Each group of questions below consists of lettered headings followed by a set of numbered items. For each numbered item select the **one** lettered heading with which it is **most** closely associated. Each lettered heading may be used **once, more than once, or not at all.**

Questions 343–344

Match the sugars below with the appropriate structure.

(A)	(B)	(C)	(D)	(E)
	CHO	CHO	CHO	CH_2OH
CHO	HCOH	HOCH	HOCH	C=O
HCOH	HOCH	HOCH	HCOH	HOCH
HCOH	HCOH	HCOH	HOCH	HCOH
HCOH	HCOH	HCOH	HCOH	HCOH
CH_2OH	CH_2OH	CH_2OH	CH_2OH	CH_2OH

343. Glucose

344. Fructose

Questions 345–348

Match the following.

(A) Low blood glucose
(B) Low blood LDL
(C) High blood cholesterol
(D) High blood glucose
(E) Low oxidation of fatty acids
(F) Ketosis
(G) Lipolysis

345. B-100 receptor deficiency

346. Result of mevinolin therapy

347. Glucose-6-phosphatase deficiency

348. Carnitine deficiency

Questions 349–350

Match each of the systematic names with the correct structural formula.

(A) $CH_3(CH_2)_{14}COOH$
(B) $CH_3(CH_2)_{16}COOH$
(C) $CH_3(CH_2)_7CH{=}CH(CH_2)_7COOH$
(D) $CH_3(CH_2)_{18}COOH$
(E) $CH_3(CH_2)_4(CH{=}CHCH_2)_4(CH_2)_2COOH$

349. Octadecanoate

350. Eicosanoate

Questions 351–354

Match each metabolic effect with the appropriate hormone.

(A) Progesterone
(B) Glucagon
(C) Aldosterone
(D) Epinephrine
(E) Thyroxine
(F) Growth hormone
(G) Insulin
(H) Glucocorticoids

351. Prevention of hypoglycemia through liver effects

352. Stimulation of gluconeogenesis

353. Stimulation of muscle glycolysis

354. Inhibition of lypolysis in adipose tissue

Questions 355–356

Match each reaction with the lettered step to which it is most closely related.

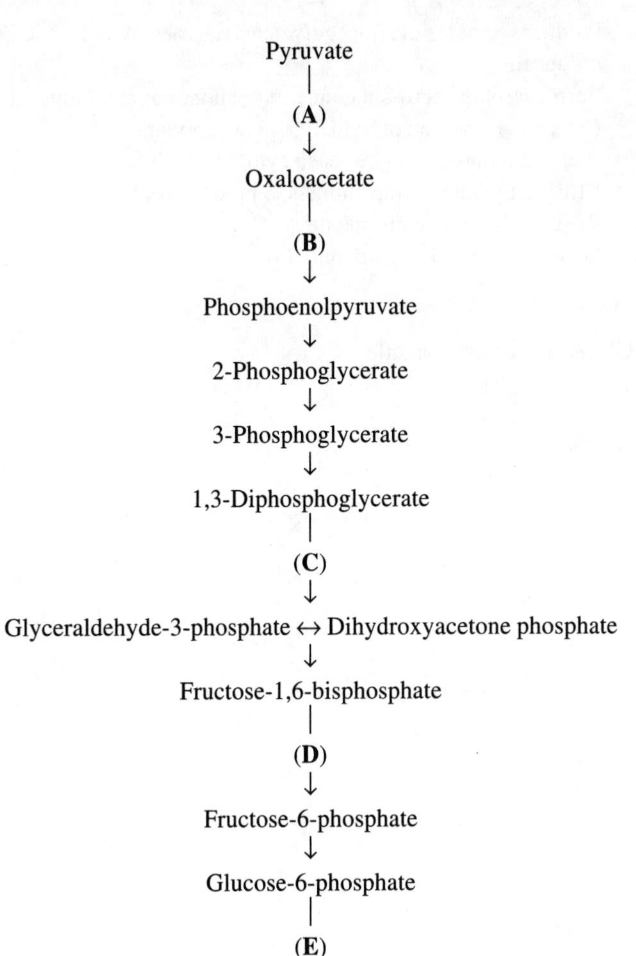

355. $ATP \rightarrow ADP + P_i$

356. $NADH \rightarrow NAD^+ + H^+$

Questions 357–360

Match each factor involved in the tricarboxylic acid cycle or oxidative phosphorylation to the appropriate description.

(A) Produces 3 moles of ATP equivalents for each cytoplasmic NADH produced
(B) Carries protons across the inner mitochondrial membrane
(C) Determines the rate of oxidative phosphorylation
(D) Yields 2 moles of ATP for each cytoplasmic NADH
(E) Utilized by hibernating animals to produce heat
(F) Requires inorganic phosphate
(G) Defines respiratory quotient (RQ)

357. Succinyl-CoA synthetase

358. Glycerol phosphate shuttle

359. Dinitrophenol

360. ADP levels

Carbohydrates and Lipids

Answers

231. The answer is B. *(Stryer, 4/e, pp 697–698.)* Chylomicra are triglyceride-rich transport particles containing dietary lipids. Very low-density lipoproteins (VLDLs) are triglyceride- and cholesterol-containing particles from the liver that contain endogenously packaged lipids. Delipidation of triglycerides from VLDLs leads to formation of intermediate forms (IDLs), and finally to a cholesterol-enriched small particle, the low-density lipoprotein (LDL). Thus, VLDL → IDL → LDL. The following table summarizes the characteristics of these plasma lipoproteins.

Type	Average Density (g/cm^3)	Percent Triglyceride	Percent Cholesterol (and Esters)
Chylomicra	0.92	85	7
VLDL	0.97	55	20
IDL		--------Intermediate--------	
LDL	1.03	10	45

232. The answer is D. *(Stryer, 4/e, pp 620–621.)* The synthesis of fatty acids requires the reduced form of nicotinamide adenine dinucleotide phosphate (NADPH) in order to reduce unsaturated carbon-carbon bonds. A major source of NADPH for fatty acid synthesis is the nicotinamide adenine dinucleotide phosphate ($NADP^+$)–linked malate enzyme. The malate enzyme is part of the citrate shuttle that transfers acetyl groups across the inner mitochondrial membrane into the cytoplasm for fatty acid synthesis. The second source of NADPH is the pentose phosphate pathway. For the synthesis of palmitate, eight NADPH molecules are formed from the citrate shuttle and six NADPH molecules from the pentose phosphate pathway. The other reduced adenine dinucleotides, nicotinamide adenine dinucleotide (NADH) and flavin adenine dinucleotide (FADH), are generated during β-oxidation of fatty acids. As a general rule, NADH and $FADH_2$ are produced in energy-yielding reactions, whereas NADPH is consumed in biosynthetic reactions.

233. The answer is C. *(Stryer, 4/e, pp 495–496.)* Glucokinase in the liver or hexokinase in other tissues catalyzes the phosphorylation of glucose as the first step of glycolysis. The equilibrium lies far to the right for the reaction as written.

$$\text{Glucse} + \text{ATP} \rightarrow \text{glucose-6-phosphate} + \text{ADP} + \text{P}_i$$

The $\Delta G°'$ for the hydrolysis of glucose-6-phosphate is −3.3 kcal/mol. Thus, the $\Delta G°'$ of the reverse reaction is +3.3 kcal/mol. Since the $\Delta G°'$ for the hydrolysis of ATP is −7.3 kcal/mol, the $\Delta G°'$ for the reaction is

$$(-7.3 \text{ kcal/mol}) + (+3.3 \text{ kcal/mol}) = -4.0 \text{ kcal/mol}$$

The phosphorylation of glucose is a thermodynamically favorable reaction.

234. The answer is B. *(Stryer, 4/e, pp 614–615.)* CO_2 is buffered in the blood into the bicarbonate system. HCO_3^- is utilized to carboxylate acetyl CoA to malonyl CoA. The regulatory enzyme acetyl CoA carboxylase catalyzes this step. The carbon contributed by HCO_3^- is immediately lost back to CO_2 when acetyl CoA is condensed with malonyl CoA in the first step of fatty acid synthesis. Thus, while a carbon from CO_2 participates in fatty acid synthesis, it does not become incorporated into fatty acids. Consequently, only one carbon of the three carbons of malonyl CoA will be labeled by radioactive CO_2.

235. The answer is B. *(Stryer, 4/e, pp 267–268.)* The most common sphingolipid in mammals is sphingomyelin. Either phosphoryl choline as shown or phosphoryl ethanolamine is the head group attached to ceramide. Ceramide, the basic structure from which all sphingolipids are derived, is composed of the 18-carbon sphingosine connected via its amino group to a fatty acid by an amide linkage. The fatty acid is usually long chain (18 to 26 carbons) and is saturated or monosaturated. Except for the lack of the glycerol backbone, sphingolipids are quite similar in structure and physical properties to the phospholipids phosphatidyl choline and phosphatidyl ethanolamine. If a neutral sugar residue is the polar head group attached to ceramide, a cerebroside is formed. If oligosaccharide head groups containing sialic acid are used, gangliosides are formed. All sphingolipids are important membrane constituents.

236. The answer is D. *(Stryer, 4/e, pp 562–563.)* As is the case with many pathways, the first step in the pentose phosphate pathway is regulated. This irreversible step is catalyzed by glucose-6-phosphate dehydrogenase, which is controlled by the level of $NADP^+$. NADPH is a competitive inhibitor of $NADP^+$. The dehydrogenation of glucose-6-phosphate produces 6-phosphoglucono-δ-lactone and NADPH.

237. The answer is E. *(Stryer, 4/e, pp 544–545.)* Inhibition of specific carriers in the electron chain produces a crossover point at the site of inhibition. Rotenone and amobarbital (Amytal) block the production of ATP at site I and the transport of electrons between NAD^+ and coenzyme Q. Thus, NADH remains reduced, while all the other elements of electron transport that feed directly into coenzyme Q are highly oxidized. Antimycin A blocks electron transport from cytochromes *b* to *c*. In this case, NADH, $FADH_2$, and everything up to and including cytochrome *b* remain reduced. Finally, cyanide blocks the passage of electrons from cytochrome a_3 to molecular oxygen at the end of electron transport. Thus, all the components of the respiratory chain remain in a highly reduced state.

238. The answer is B. *(Stryer, 4/e, p 553.)* Although metabolic poisons do uncouple oxidation of NADH from production of ATP, the uncoupling of oxidative phosphorylation does occur under certain normal biologic conditions. In brown fat (so-called because of the large number of blood vessels), oxidation is uncoupled from phosphorylation so that heat is produced. This phenomenon is observed in newborn mammals, including humans, hibernating animals, and certain mammals that use it as an adaptation to the cold. In brown fat, fatty acids act as uncouplers of mitochondria. The catecholamine norepinephrine controls the release of the fatty acids, thereby imparting regulation of the process.

239. The answer is C. *(Stryer, 4/e, pp 624–625.)* During the synthesis of prostaglandins, a specific fatty acid is released from the 2′ position of membrane phospholipids by the action of phospholipase A_2. After its release, the fatty acid can enter either the lipoxygenase pathway, which produces acid with an unknown biologic function, or the prostaglandin cyclooxygenase (also called *prostaglandin synthetase*) pathway. In the formation of prostaglandins from fatty acids, cyclooxygenase catalyzes formation of a cyclopentane ring and the introduction of three oxygen atoms. The type of prostaglandin produced depends on the starting fatty acid, which is always a derivative of an essential fatty acid. Eicosatrienoic acid yields series 1 prostaglandins, eicosatetraenoic (arachidonic) acid yields series 2 prostaglandins, and eicosapentaenoic acid yields series 3 prostaglandins. Aspirin, as well as indomethacin, decreases prostaglandin synthesis by inhibiting the oxygenase activity of cyclooxygenase.

240. The answer is D. *(Stryer, 4/e, pp 696–697, 701–702.)* Regulation of cholesterol metabolism is by definition exerted at the "committed" and rate-controlling step. This is the reaction catalyzed by 3-hydroxy-3-methylglutaryl CoA reductase. Reductase activity is reduced by fasting and by cholesterol feeding and thus provides effective feedback control of cholesterol metabolism.

241. The answer is C. *(Stryer, 4/e, pp 534–541.)* Many different substrates donate electrons to NAD^+ to produce NADH. In electron transport, NADH denotes electrons to flavin mononucleotide (FMN). Alternatively, some substrates are dehydrogenated by flavin-linked dehydrogenases and skip NAD^+. In any event, all these electrons flow through coenzyme Q and then through the cytochromes. The correct order is $b \rightarrow c_1 \rightarrow c \rightarrow a \rightarrow a_3$.

242. The answer is A. *(Stryer, 4/e, pp 466–468.)* The molecule depicted in the question is α-D-glucopyranose. It is one of a series of D-glucose hemiacetals in which there are alternate α and β forms available because of the asymmetry in the terminal carbon. Some of the glycosides from the β form, such as the steroid-containing cardiac glycosides, are pharmacologically important.

243. The answer is B. *(Stryer, 4/e, pp 496–498.)* In humans, ethanol is cleared from the body by oxidation catalyzed by two NAD^+-linked enzymes: alcohol dehydrogenase and acetaldehyde dehydrogenase. These enzymes act mainly in the liver to convert alcohol to acetaldehyde and acetate, respectively. In chronic alcoholics, alcohol dehydrogenase may be elevated somewhat. The NADH level is significantly increased in the liver during oxidation of alcohol, owing to the consumption of NAD^+. This leads to a swamping of the normal means of regenerating NAD^+. Thus, NAD^+ becomes the rate-limiting factor in oxidation of excess alcohol.

244. The answer is E. *(Stryer, 4/e, pp 267–269.)* Choline is the alcoholic amine that, together with phosphate, forms the hydrophilic unit of sphingomyelin. Phosphorylcholine plays a similar structural part as a polar head group in the phospholipid phosphatidylcholine. The hydrophilic unit of phosphatidic acid is a phosphorylated alcohol. Glucose is the hydrophilic moiety in glucocerebroside, whereas the OH group at C_3 is the only hydrophilic group of cholesterol.

245. The answer is C. *(Stryer, 4/e, pp 622–623.)* In mammals, arachidonic (5,8,11,15-eicosatetraenoic) acid can only be synthesized from essential fatty acids derived from the diet. Linoleic (9,12-octadecadienoic) acid will produce arachidonic acid following two desaturations and chain elongation. While linolenic (9,12,15-octadecatrienoic) acid also is an essential fatty acid, desaturation and elongation would produce 8,11,14,17-eicosatetraenoic acid, which is distinct from arachidonic acid. Oleic, palmitic, and stearic acids are all non-essential fatty acids that cannot give rise to arachidonic acids in mammals.

246. The answer is C. *(Stryer, 4/e, pp 245–246.)* Cyclic AMP is synthesized by adenylate cyclase in response to hormonal stimulation of specific receptors in cells. In all eukaryotic cells studied to date, increased cyclic AMP levels activate a cyclic AMP–dependent protein kinase. The protein kinase, in turn, phosphorylates other enzymes, activating or inactivating them. The kinase is made up of two regulatory and two catalytic subunits. Binding of cyclic AMP to the regulatory subunits allows dissociation of the catalytic subunits. The catalytic subunits are active in the dissociated state. Decrease in cellular cyclic AMP levels by phosphodiesterase breakdown of the nucleotide frees the regulatory subunits from cyclic AMP binding. This allows reassociation of regulatory and catalytic subunits and subsequent inactivation.

247. The answer is D. *(Stryer, 4/e, pp 544–545.)* Cyanides inhibit the action of cytochrome oxidase, a key enzyme in the process of tissue respiration. In so doing, they reduce respiration by 60 to 90 percent, which indicates that this proportion of animal cell respiration involves the electron transport system terminating in cytochrome oxidase.

248. The answer is B. *(Stryer, 4/e, pp 544–545.)* The chemiosmotic hypothesis of Mitchell describes the coupling of oxidative phosphorylation and electron transport. The movement of electrons along the electron transport chain allows protons to be pumped from the matrix of the mitochondria to the cytoplasmic side. The protons are pumped at three sites in the electron transport chain to produce a proton gradient. When protons flow back through proton channels of the asymmetrically oriented ATPase of the inner mitochondrial membrane, ATP is synthesized. It is now generally accepted that the Mitchell hypothesis correctly accounts for the connection between oxidative phosphorylation and electron transport.

249. The answer is C. *(Stryer, 4/e, pp 685–686.)* In order for the synthesis of triacylglycerol to occur, glycerol and fatty acids must be activated. Glycerol is activated by being phosphorylated to glycerol-3-phosphate by glycerol kinase. ATP contributes one phosphate. Free fatty acids are activated by forming thioesters with CoA to become fatty acyl CoA. Fatty acyl-CoA synthetase (thiokinase) catalyzes this reaction. It is energized by the breakdown of ATP to AMP + PP_i. Thus, the equivalent of two high-energy phosphate bonds is used per fatty acid activated. Since three fatty acids are esterified to one glycerol for each triacylglycerol synthesized, 7 moles of high-energy phosphate bond equivalents are used.

250. The answer is D. *(Stryer, 4/e, p 767.)* A facultative cell, which can metabolize glucose under anaerobic or aerobic conditions, that is switched from anaerobic to aerobic metabolism experiences a sharp decrease in rate of glucose utilization. Under aerobic conditions, the cell can produce a net gain in moles of ATP formed per mole of glucose utilized that can be as high as 18 times that produced under anaerobic conditions. Thus the cell requires less glucose. Such increased ATP concentrations, together with the release of citrate from the citric acid cycle under aerobic conditions, allosterically inhibit the key regulatory enzyme of the glycolytic pathway, phosphofructokinase.

251. The answer is D. *(Stryer, 4/e, p 489.)* Even in low concentration, fluoride is an inhibitor of enolase, an enzyme involved in the Embden-Meyerhof pathway of glycolysis. Fluoride combines with magnesium, calcium, and other divalent metals in the enzymes it affects, possibly by means of a fluorophosphate linkage.

252. The answer is B. *(Stryer, 4/e, pp 490–491.)* Aerobic glycolysis can be defined as the oxidative conversion of glucose to two molecules of pyruvate. In the process, two molecules of ATP and two molecules of NADH are produced. Since reducing equivalents from the two molecules of NADH produced in the cytoplasm must be transported into the mitochondrion for oxidation, it is not known how many ATP molecules are produced. On the assumption that two ATP molecules are formed per molecule of NADH oxidized via the glycerol phosphate shuttle, the ATP yield in aerobic glycolysis can be calculated as six ATP molecules per mole of glucose utilized.

253. The answer is B. *(Stryer, 4/e, pp 511–512.)* A molecule of guanosine triphosphate is synthesized from guanosine diphosphate and phosphate at the cost of hydrolyzing succinyl CoA to succinate and CoA. This constitutes substrate-level phosphorylation, and in contrast to oxidative phosphorylation, this is the only reaction in the citric acid cycle that directly yields a high-energy phosphate bond. The sequence of reactions from α-ketoglutarate to succinate is catalyzed by the α-ketoglutarate dehydrogenase complex and succinyl-CoA synthetase, respectively.

$$\alpha\text{-ketoglutarate} + NAD^+ + CoA \rightarrow \text{succinyl CoA} + CO_2 + NADH$$
$$\text{succinyl CoA} + P_i + GDP \rightarrow \text{succinate} + GTP + CoA$$

254. The answer is E. *(Stryer, 4/e, pp 618–620.)* The fatty acid synthase complex of mammals is composed of two identical subunits. Each of the subunits is a multienzyme complex of seven enzymes and the acyl carrier protein component. All the components are covalently linked together; thus, all the components are on a single polypeptide chain, which functions in the presence of another identical polypeptide chain. Each cycle of fatty acid synthesis employs the acyl carrier protein and six enzymes: acetyl transferase, malonyl transferase, β-ketoacyl synthase, β-ketoacyl reductase, dehydratase, and enoyl reductase. When the final fatty acid length is reached (usually C_{16}), thioesterase hydrolyzes the fatty acid off of the synthase complex.

255. The answer is C. *(Stryer, 4/e, pp 447–449.)* The two glycolytic reactions involving respectively an oxidation and a phosphate transfer, in which glyceraldehyde-3-phosphate is converted to 3-diphosphoglycerate, yield an overall energy change given by the expression

$$\Delta G^{\circ\prime} \text{ total} = \Delta G^{\circ\prime}_1 + \Delta G^{\circ\prime}_2 = +1.5 + (-4.5) = -3.0 \text{ kcal/mol}$$

256. The answer is E. *(Stryer, 4/e, pp 570–573.)* The enzyme controlling the first step in gluconeogenesis is pyruvate carboxylase. It catalyzes the conversion of pyruvate to oxaloacetate. Pyruvate is absolutely dependent upon the presence of the allosteric effector acetyl CoA or a closely related acyl CoA for its function. Under conditions of high energy charge and high levels of acetyl CoA, oxaloacetate is utilized for gluconeogenesis. If low amounts of ATP are present, oxaloacetate is consumed in the citric acid cycle.

257. The answer is C. *(Stryer, 4/e, pp 490–491.)* Fructose-1,6-diphosphate, one of the intermediaries in glycolysis, is synthesized from fructose-6-phosphate by phosphofructokinase; it constitutes the "committed step" in the glycolytic process. Aldolase catalyzes the reversible interconversion between fructose-1,6-diphosphate on the one hand and glyceraldehyde-3-phosphate and dihydroxyacetone phosphate on the other.

258. The answer is D. *(Stryer, 4/e, pp 607–608.)* Long-chain fatty acids activated in the cytoplasm are carried across the inner mitochondrial membrane by carnitine. The fatty acyl group is transferred from CoA to carnitine, traverses the membrane, and then is transferred back to CoA. Carnitine acyltransferase I on the cytolic side of the inner mitochondrial membrane and carnitine acyltransferase II on the matrix face catalyze this reaction. Medium-chain fatty acids do not require carnitine to diffuse across either the inner mitochondrial membrane or plasma membranes.

259. The answer is B. *(Stryer, 4/e, p 472.)* In many populations, a majority of adults are deficient in lactase and hence intolerant to the lactose in milk. In all populations, at least some adults have this deficiency. Since virtually all children are able to digest lactose, this deficiency develops in adulthood. In such lactase-deficient adults, lactose accumulates in the small intestine because no transports exist for the disaccharide. An outflow of water into the gut owing to the osmotic effect of the milk sugar causes the clinical symptoms. Steatorrhea, or fatty stools, is caused by unabsorbed fat, which can occur following a fatty meal in persons with a deficiency of lipoprotein lipase.

260. The answer is E. *(Stryer, 4/e, pp 700–702.)* In normal persons, plasma cholesterol levels average about 175 mg/dL. The level in heterozygotes with the autosomal dominant genotype is about 300 mg/dL and in homozygotes about 680 mg/dL. The single mutagenic defect results in a functional loss of low-density lipoprotein (LDL) receptors on the plasma membranes of cells other than those of the liver. This prevents the normal clearing of LDLs from the blood plasma by endocytosis.

261. The answer is E. *(Stryer, 4/e, pp 585–586.)* The following reaction,

$$\text{UDP glucose} + \text{glycogen}_n \rightarrow \text{UDP} + \text{glycogen}_{n+1}$$

which is reversible, is catalyzed by UDP-sugar phosphorylases (uridyltransferases). The nucleotide uridine diphosphate, in this and other ways, is closely related to carbohydrate metabolism.

262. The answer is A. *(Stryer, 4/e, pp 485–487.)* In the following reaction

$$\text{ATP} + \text{glucose} \rightarrow \text{glucose-6-phosphate} + \text{ADP}$$

ATP is the high-energy phosphate employed for phosphorylation of glucose via hexokinase or glucokinase. It is used also to phosphorylate fructose and fructose-6-phosphate.

263. The answer is D. *(Stryer, 4/e, pp 622–624.)* Most mammals have the ability to synthesize nonessential fatty acids and cholesterol from acetyl CoA. Phospholipids, sphingolipids, and triacylglycerols, which utilize fatty acids, can also be synthesized de novo. However, essential fatty acids will be missing from animals raised on a fat-free diet. Linoleate (18:2 *cis*-9,12), linolenate (18:3 *cis*-9,12,15), and their derivatives can only be obtained by animals from plants or the fat depots of other animals that have eaten plants. All the prostaglandins are synthesized from essential fatty acids. Prostaglandins are potent hormonelike substances produced by most tissues.

264. The answer is E. *(Stryer, 4/e, pp 612–613, 693–695.)* The major fate of acetoacetyl CoA formed from condensation of acetyl CoA in the liver is the formation of 3-hydroxy-3-methylglutaryl CoA (HMG CoA). Under normal postabsorptive conditions, HMG CoA production occurs in the cytoplasm of hepatocytes as part of the overall process of cholesterol biosynthesis. However, in fasting or starving persons, as well as in patients with uncontrolled diabetes mellitus, HMG CoA production occurs in liver mitochondria as part of ketone body synthesis. In this process, HMG CoA is cleaved by HMG CoA lyase to yield acetoacetate and acetyl CoA. The NADH-dependent enzyme β-hydroxybutyrate dehydrogenase converts most of the acetoacetate to β-hydroxybutyrate. These two ketone bodies, acetoacetate and β-hydroxybutyrate, diffuse into the blood and are transported to peripheral tissues.

265. The answer is C. *(Stryer, 4/e, pp 496–498.)* During the fermentation of grapes or other plant products, yeast converts pyruvate to ethanol. This is a two-step reaction. The first step is the formation of acetaldehyde from pyruvate catalyzed by pyruvate decarboxylase.

$$\text{Pyruvate} + H^+ \rightarrow \text{acetaldehyde} + CO_2$$

In the second step, alcohol dehydrogenase catalyzes the reduction of acetaldehyde to ethanol and the regeneration of NAD^+ from NADH.

$$\text{Acetaldehyde} + \text{NADH} + H^+ \rightarrow \text{ethanol} + NAD^+$$

This last step regenerates NAD^+ so that glycolysis may continue. In resting muscle, almost all pyruvate produced is converted to acetyl CoA for oxidation in the citric acid cycle. In actively contracting muscle, when oxygen is limited, lactate accumulates so that NAD^+ is regenerated. Acetyl CoA, citrate, and oxaloacetate can all be found in muscle-tissue mitochondria.

266. The answer is B. *(Stryer, 4/e, pp 539–541.)* Cytochromes *a* and a_3 exist as a complex known as *cytochrome oxidase.* They are the last members of the electron transport chain. They transfer electrons between cytochrome *c* and O_2. Four electrons and four protons are required for each molecular oxygen (O_2) reduced to water ($2H_2O$). Copper is involved in the electron transfers of the oxidase.

267. The answer is E. *(Stryer, 4/e, p 696.)* Bile acids often are conjugated with glycine to form glycocholic acid and with taurine to form taurocholic acid. In human bile, glycocholic acid is by far the more common. The presence of the charged carboxyl group of glycine or the charged sulfate of taurine adds to the hydrophilic nature of the bile acids, thereby increasing their ability to emulsify lipids during the digestive process.

268. The answer is E. *(Stryer, 4/e, pp 489–491.)* ATP is synthesized by two reactions in glycolysis: the reactions catalyzed by phosphoglycerate kinase (reaction E-F in the pathway exhibited in the question) and pyruvate kinase (reaction G-H). Compound A is glucose-6-phosphate; B is fructose-6-phosphate; C is fructose-1,6-diphosphate; D is glyceraldehyde-3-phosphate; E is 1,3-diphosphoglycerate; F is 3-phosphoglycerate; G is phosphoenolpyruvate; and H is pyruvate.

269. The answer is E. *(Stryer, 4/e, pp 622–623.)* Linoleic and linolenic fatty acids cannot be synthesized by mammals and are, therefore, referred to as *essential.* Other fatty acids—including palmitic, stearic, and oleic acids—that do not contain double bonds between the seventh carbon from the terminal methyl group and the carboxyl group can be made by alternate desaturation and elongation in mammals.

270. The answer is B. *(Stryer, 4/e, pp 577–578.)* The end product of glycolysis in erythrocytes, which have no mitochondria, is lactate. The conversion of pyruvate to lactate by lactate dehydrogenase allows a regeneration of NAD^+ from the NADH accumulated in the oxidation of glyceraldehyde-3-phosphate. This allows glycolysis to proceed. Lactate also accumulates in muscle during heavy exercise, when oxygen becomes limiting. In mammals, the lactate diffuses into the bloodstream and is metabolized by the liver. A similar formation of lactate occurs in many microorganisms when they are grown under anaerobic conditions. Alternatively, ethanol may be formed by microorganisms (but not erythrocytes) when they are faced with anaerobic conditions. In all cases, the end products formed allow regeneration of NAD^+ to occur and glycolysis to proceed.

271. The answer is C. *(Stryer, 4/e, pp 496–498.)* The principal pathway for hepatic metabolism of ethanol is thought to be oxidation to acetaldehyde in the cytoplasm by alcohol dehydrogenase. Acetaldehyde is then oxidized, probably within the mitochondrion, to yield acetate. Acetone, methanol, hydrogen peroxide, and glycerol do not appear in this biodegradation pathway.

272. The answer is D. *(Stryer, 4/e, pp 697–698.)* The shell of apoproteins coating blood transport lipoproteins is important in the physiologic function of the lipoproteins. Some of the apoproteins contain signals that target the movement of the lipoproteins in and out of specific tissues. B-48 and E seem to be important in targeting chylomicron remnants to be taken up by liver. B-100 is synthesized as the coat protein of VLDLs and marks their end product, LDLs, for uptake by peripheral tissues. Other apoproteins are important for the solubilization and movement of lipids and cholesterol in and out of the particles. C-II is a lipoprotein lipase activator that VLDLs and chylomicrons receive from HDLs. The A apoproteins are found in HDLs and are involved in lecithin-cholesterol acyl transferase (LCAT) regulation.

273. The answer is E. *(Stryer, 4/e, pp 490–491.)* High-energy phosphate bonds are added to the substrates of glycolysis at three steps in the pathway. Hexokinase, or in the case of liver, glucokinase, adds phosphate from ATP to glucose to form glucose-6-phosphate. Strictly speaking, this is not always considered a step of the glycolytic pathway. Phosphofructokinase uses ATP to convert fructose-6-phosphate to fructose-1,6-phosphate. Using NAD^+ in an oxidation-reduction reaction, inorganic phosphate is added to glyceraldehyde-3-phosphate by the enzyme glyceraldehyde-3-phosphate dehydrogenase to form 1,3-diphosphoglycerate. The enzymes phosphoglycerate kinase and pyruvate kinase transfer substrate high-energy phosphate groups to ADP to form ATP.

274. The answer is A. *(Stryer, 4/e, pp 491–492.)* Fructose is taken in by humans both as a part of sucrose and as a free sugar. During its catabolism, fructose is first phosphorylated to fructose-1-phosphate by fructokinase. Aldol cleavage by fructose-1-phosphate–specific aldolase, not enolase, yields glyceraldehyde and dihydroxyacetone phosphate. The glyceraldehyde is phosphorylated to glyceraldehyde-3-phosphate by triose kinase. Consequently, both triose phosphates enter glycolysis. In adipocytes, fructose can be alternatively phosphorylated by hexokinase to fructose-6-phosphate. However, this reaction will be competitively inhibited by appreciable amounts of glucose as it is in other tissues.

275. The answer is D. *(Stryer, 4/e, pp 490–491.)* The conversion of glucose-6-phosphate to fructose-6-phosphate is catalyzed by phosphoglucose isomerase. In this isomerization, a six-membered pyranose ring is converted to a five-membered furanose ring; that is, an aldose is converted into a ketose. The isomerization of dihydroxyacetone phosphate to glyceraldehyde-3-phosphate is a conversion of a ketose into an aldose. This isomerization by triose phosphate isomerase allows dihydroxyacetone phosphate to reenter the glycolytic pathway, since the next enzymatic step is specific only for glyceraldehyde-3-phosphate. The conversion of 3-phosphoglycerate to 2-phosphoglycerate is a phosphoryl shift catalyzed by phosphoglyceromutase. The formation of phosphoenolpyruvate from 2-phosphoglycerate is a dehydration reaction catalyzed by enolase. ATP is required for the phosphorylation of glucose to glucose-6-phosphate and ATP is produced from ADP when 1,3-bisphosphoglycerate is dephosphorylated to produce 3-phosphoglycerate.

276. The answer is B. *(Stryer, 4/e, pp 696–698.)* In the postabsorptional (postprandial) state, plasma contains all the lipoproteins: chylomicra derived from dietary lipids packaged in the intestinal epithelial cells and their remnants; very low-density lipoproteins (VLDLs), which contain endogenous lipids and cholesterol packaged in the liver; low-density lipoproteins (LDLs), which are end products of delipidation of VLDLs; and high-density lipoprotins (HDLs), which are synthesized in the liver. HDLs are, in part, catalytic, since transfer of their C-II apolipoprotein to VLDLs or chylomicra activates lipoprotein lipase. Following a 12-h fast in normal patients, only LDLs and HDLs remain in plasma, since both chylomicra and VLDLs have been delipidated. Most of the cholesterol measured in blood plasma at this time is present in the cholesterol-rich LDLs. However, HDL cholesterol also contributes to the measurement.

277. The answer is A. *(Stryer, 4/e, pp 612, 641.)* Lipolysis of triacylglycerols yields fatty acids and glycerol. The free glycerol is transported to the liver, where it can be phosphorylated to glycerol phosphate and enter the glycolysis or the gluconeogenesis pathways at the level of dihydroxyacetone phosphate. Acetyl CoA and propionyl CoA are produced in the final round of degradation of an odd-chain fatty acid. Acetyl CoA cannot be converted to glucose, but propionyl CoA can. The three carbons of propionyl CoA enter the citric acid cycle after being converted into succinyl CoA. Then, succinyl CoA can be converted to oxaloacetate and enter the glycolytic scheme. Ketone bodies, including β-hydroxybutyrate, are produced from acetyl CoA units derived from fatty acid β-oxidation. They may not be converted to glucose. Amino acids are not a product of triacylglycerol breakdown.

278. The answer is B. *(Stryer, 4/e, pp 493–494, 621–622.)* Under conditions in which the entry charge of liver cells is high, intermediates of the citric acid cycle are abundant. Citrate, an early intermediate in the cycle, readily diffuses across the inner membrane of mitochondria and out into the cytosol.

Citrate allosterically inhibits phosphofructokinase and, conversely, stimulates fructose-1,6-diphosphatase. Thus, when the energy level of hepatocytes is low and biosynthetic precursors are not abundant, phosphofructokinase is stimulated and glycolysis is favored. When the energy level is high, citrate inhibits phosphofructokinase, stimulates the diphosphatase, and thereby promotes gluconeogenesis. The diffusion of high levels of citrate into the cytosol also stimulates synthesis of fatty acids. Citrate activates acetyl CoA carboxylase, the first step in the synthesis of fatty acids, as well as provides its substrates, acetyl CoA and NADPH. However, citrate does not allosterically activate fatty acid synthetase. Enolase, an enzyme of the glycolytic pathway, is not regulated.

279. The answer is D. *(Stryer, 4/e, pp 570–572.)* All the enzymes listed are specific to either glycolysis or gluconeogenesis, except for phosphoglycerate kinase. It is one of seven enzymes common to both glycolysis and gluconeogenesis. The enzymes hexokinase, phosphofructokinase, and pyruvate kinase catalyze irreversible reactions unique to glycolysis. In order for gluconeogenesis to occur, the three irreversible reactions must be replaced. Pyruvate is synthesized into phosphoenolpyruvate by a two-step reaction. First, oxaloactate is formed by carboxylation in the presence of pyruvate carboxylase. Then, phosphoenolpyruvate carboxykinase decarboxylates and phosphorylates oxaloacetate in the presence of GTP. The next irreversible step to be bypassed in gluconeogenesis requires fructose-6-phosphate to be produced by the action of fructose-1,6-phosphatase on fructose-1,6-phosphate. When glucose-6-phosphate is finally produced during gluconeogenesis, it is converted to glucose by glucose-6-phosphatase, an enzyme unique to the endoplasmic reticulum. The free glucose may then diffuse from the liver into the bloodstream. Of the enzymes given as possible answers, only phosphoglycerate kinase is an enzyme that catalyzes a reversible reaction common to both glycolysis and gluconeogenesis.

280. The answer is B. *(Stryer, 4/e, pp 605–606.)* Lipolysis is directly regulated by hormones in adipocytes. Epinephrine stimulates adenylate cyclase to produce cyclic AMP, which in turn stimulates a protein kinase. The kinase activates triglyceride lipase by phosphorylating it. Lipolysis then proceeds and results in the release of free fatty acids and glycerol. A futile reesterification of free fatty acids is prevented, since adipocytes have little glycerol kinase to phosphorylate the liberated glycerol, which must be processed in the liver. Inhibition of lipolysis occurs in the presence of insulin, which lowers cyclic AMP levels. Lipoprotein lipase is not an adipocyte enzyme.

281. The answer is D. *(Stryer, 4/e, pp 606–607.)* Fatty acids must be activated before being oxidized. In this process, they are linked to CoA in a reaction catalyzed by thiokinase (also known as *acyl-CoA synthetase*). ATP is hydrolyzed to AMP plus pyrophosphate in this reaction. In contrast, the enzyme thiolase cleaves off acetyl CoA units from β-ketoacyl CoA, while it forms thioesters during β-oxidation.

282. The answer is D. *(Stryer, 4/e, pp 544–548.)* Oligomycin inhibits mitochondrial ATPase and thus prevents phosphorylation of ADP to ATP. It prevents utilization of energy derived from electron transport for the synthesis of ATP. Oligomycin has no effect on coupling but blocks mitochondrial phosphorylation so that both oxidation and phosphorylation cease in its presence.

283. The answer is E. *(Stryer, 4/e, pp 588–589.)* Glycogen is an energy-storage molecule mainly composed of glucose residues linked in α-1,4-glycosidic residues. These residues are sequentially cleaved off during hydrolysis by glycogen phosphorylase in the presence of inorganic phosphate to yield glucose-1-phosphate. The glucose-1-phosphate is converted to glucose-6-phosphate by phosphoglucomutase. Thus, phosphorylated glucose is derived from glycogen at no relative cost.

284. The answer is A. *(Stryer, 4/e, p 691.)* Gangliosides are continually synthesized and broken down. The specific hydrolases that degrade gangliosides by sequentially removing terminal sugars are found in lysosomes. In the lipid-storage disease known as *Tay-Sachs disease,* ganglioside G_{M2} accumulates because of a deficiency of β-*N*-acetylhexosaminidase, a lysosomal enzyme that removes the terminal *N*-acetylgalactosamine residue. Homozygotes produce virtually no functional enzyme and suffer weakness, retardation, and blindness. Death usually occurs before infants are 3 years old. Carriers (heterozygotes) of the autosomal recessive disease produce approximately 50 percent of the normal levels of enzyme but show no ill effects.

285. The answer is B. *(Stryer, 4/e, pp 513–514.)* Reducing equivalents are produced at four sites in the citric acid cycle. NADH is produced by the isocitrate dehydrogenase–catalyzed conversion of α-ketoglutarate to succinyl CoA, and the malate dehydrogenase–catalyzed conversion of malate to oxaloacetate. $FADH_2$ is produced by the succinate dehydrogenase–catalyzed conversion of succinate to fumarate. Succinyl CoA synthetase catalyzes the formation of succinate from succinyl CoA, with the concomitant phosphorylation of GDP to GTP.

286. The answer is C. *(Stryer, 4/e, pp 587–588, 594–596.)* Glycogen synthetase is an enzyme that transfers glucosyl moieties from UDP-glucose to a glycogen polymer primer. In plants, ADP-glucose plays a part similar to that of UDP-glucose in animals. The enzyme exists in two forms: an active, dephosphorylated form; and an inactive, phosphorylated form. It is inactivated by phosphorylation of a specific serine residue. Glycogen breakdown, not synthesis, is positively affected by increased calcium levels.

287. The answer is E. *(Stryer, 4/e, pp 494–495, 770–773.)* The main control of glycolysis is through the enzyme phosphofructokinase. This enzyme is controlled by a high level of ATP, which inhibits it, or a high level of fructose-2,6-bisphosphate (F-2,6-BP), which activates it. The inhibitory effect of ATP is potentiated by citrate, while high AMP levels reverse it. During fasting, when blood glucose levels are low, a glucagon-signaled increase of liver cyclic AMP leads to the activation of a phosphatase that hydrolyzes the 2-phosphoryl group from F-2,6-BP. The same glucagon-stimulated cascade deactivates the kinase that phosphorylates fructose-6-phosphate. The subsequent lowering of F-2,6-BP inactivates phosphofructokinase.

288. The answer is A. *(Stryer, 4/e, pp 620–621.)* The sources of NADPH for synthesis of fatty acids are the pentose phosphate pathway and cytosolic malate formed during the transfer of acetyl groups to the cytosol as citrate. The enzyme citrate lyase splits citrate into acetyl CoA and oxaloacetate. The oxaloacetate is reduced to malate by NADH. NADP-linked malate enzyme catalyzes the oxidative decarboxylation of malate to pyruvate and carbon dioxide. Thus, the diffusion of excess citrate from the mitochondria to the cytoplasm of cells not only provides acetyl CoA for synthesis of fatty acids but NADPH as well. One NADPH is produced for each acetyl CoA produced. However, most of the NADPHs needed for synthesis of fatty acids are derived from the pentose phosphate pathway. For this reason, adipose tissue has an extremely active pentose phosphate pathway.

289. The answer is E. *(Stryer, 4/e, pp 612, 638, 641.)* The final thiolytic cleavage in β-oxidation of odd-chain fatty acids yields propionyl CoA. Propionyl CoA is also formed during the breakdown of methionine and isoleucine. It is carboxylated to form D-methylmalonyl CoA, which is in equilibrium with L-methylmalonyl CoA. Valine forms methylmalonyl CoA during its degradation. The L-isomer of methylmalonyl CoA is converted to succinyl CoA through the action of the B_{12} coenzyme–containing methylmalonyl CoA mutase. Thus, succinyl CoA serves as the entry point into the citric acid cycle for three amino acids and the last three carbons of odd-chain fatty acids. The amino acids and fatty acid carbons introduced in this manner may either be catabolized in the cycle for energy production or utilized for gluconeogenesis.

290. The answer is A. *(Stryer, 4/e, pp 513–514.)* Given an excess of acetyl CoA, the soluble components of the cycle would run until all the NAD^+ and FAD cofactors were reduced. Since components of the electron transport chain were not specified as being present, these cofactors cannot be reoxidized. Molecular oxygen is not used directly by the citric acid cycle but is required by the electron transport chain. The intermediates of the cycle need only be present in trace amounts since they are not consumed. Likewise, the enzymes of the cycle would still be functional when the cofactors were reduced.

291. The answer is D. *(Stryer, 4/e, pp 493–495, 766–768.)* Epinephrine stimulates both muscle and liver adenylate cyclase to produce cyclic AMP. In liver, the increased cyclic AMP levels activate a phosphatase that dephosphorylates fructose-2,6-bisphosphate (F-2,6-BP), while deactivating a kinase that produces F-2,6-BP. Thus, F-2,6-BP levels are decreased and phosphofructokinase activity is decreased. In liver and muscle, F-2,6-BP is the major allosteric activator of phosphofructokinase. In skeletal muscle, however, the kinase responsible for the synthesis of F-2,6-BP is activated, not inhibited, by cyclic AMP. Thus, muscle sees an increase in glycolysis following epinephrine stimulation, while liver experiences a decrease in glycolytic activity. In both tissues, glycogen phosphorylase is activated and glycogenolysis occurs. Under these conditions, glucose is utilized in muscle for ATP production relative to contractile activity, while liver produces glucose for export to the blood.

292. The answer is A. *(Stryer, 4/e, pp 611–612.)* β-Oxidation accounts for most of the catabolism of fatty acids. However, branched-chain fatty acids cannot be broken down by β-oxidation when a methyl group is found on the β-carbon. Phytanic acid, a plant branched-chain fatty acid, requires α-oxidation. This is a process whereby the carboxyl group of fatty acids is removed as CO_2 when a β-branch methyl is encountered. Propionic acid is removed in the next step, which eliminates the branch. Phytanic acid accumulates in persons suffering from Refsum's disease, a deficiency of α-oxidation. ω-Oxidation is the oxidation of terminal methyl groups from fatty acids to yield dicarboxylic acids.

293. The answer is B. *(Stryer, 4/e, pp 685–686.)* Triacylglycerides are synthesized from free fatty acids and glycerol phosphate in adipocytes and most other cells. In these cells, glycerol phosphate is converted to 3′-diacylglyceride, which then interacts with fatty acyl CoA to produce triacylglyceride. In contrast, in the endoplasmic reticulum of intestinal epithelial cells, 2′-diacylglycerides derived from digestion are combined with fatty acyl CoA to yield triacylglycerides. This process represents the reesterification of triacylglycerides from fatty acids and 2′-diacylglycerides formed by the action of pancreatic lipase in the gut lumen. The triacylglycerides produced by the gut epithelium are packaged into chylomicrons for transport. Acyl-carrier protein is found in fatty acid synthase and is required for fatty acid synthesis.

294. The answer is A. *(Stryer, 4/e, pp 491–492.)* While most tissues cannot utilize fructose, liver, kidneys, intestine, and adipose tissue can. Genetic fructokinase deficiency causes no symptoms. It can be detected by urine measurements of fructose that spills over into the urine. Unless care is taken, this could be misinterpreted as glucosuria, like that seen in diabetes, since both fructose and glucose are positive for a reducing-sugar test. Liver hexokinase rarely phosphorylates fructose to fructose-6-phosphate because the liver enzyme has a much greater affinity for glucose. However, adipose tissue hexokinase produces fructose-6-phosphate, which then can be acted upon by fructose-1-phosphate aldolase, which splits it into hydroxyacetone phosphate and glyceraldehyde. Glyceraldehyde and dihydroxyacetone phosphate proceed through glycolysis or gluconeogenesis through the action of triose kinase. Under normal circumstances, liver fructokinase phosphorylates fructose to fructose-1-phosphate, and fructose-1-phosphate aldolase acts upon it.

295. The answer is B. *(Stryer, 4/e, pp 491–493.)* Under ordinary circumstances, galactose is quickly converted into glucose-6-phosphate by the liver or erythrocytes. Most other organs do not metabolize galactose. First, galactose is phosphorylated by ATP to produce galactose-1-phosphate in the presence of galactokinase. Next, galactose-1-phosphate uridyl transferase transfers UDP from UDP-glucose to form UDP-galactose and glucose-1-phosphate. Under the action of UDP-galactose-4-epimerase, UDP-galactose is epimerized to UDP-glucose. Finally, glucose-1-phosphate is isomerized to glucose-6-phosphate by phosphoglucomutase. The genetic deficiency of galactose-1-phosphate uridyl transferase causes galactosemia. In this autosomal recessive disease, affected infants do not do well unless milk is excluded from the diet, thereby eliminating lactose. Vomiting, diarrhea, enlargement of the liver, jaundice, and ultimately mental retardation occur unless milk is eliminated from the diet. The developmental problems are caused by the reduction of galactose to galactitol (dulcitol) in the presence of the enzyme aldose reductase. High levels of galactitol cause cataracts. In addition, an accumulation of galactose-1-phosphate causes liver enlargement.

296. The answer is C. *(Stryer, 4/e, pp 605–606.)* During lipolysis, triglycerides are split into three free fatty acids and glycerol. The free fatty acids, as well as the free glycerol, diffuse into the bloodstream where they are circulated throughout the body. The free fatty acids are used as an energy source for many tissues, primarily muscle. The free glycerol that is released cannot be phosphorylated back to glycerol-3-phosphate in the adipose tissue since it lacks glycerol kinase. However, the free glycerol released in lipolysis is taken up by the liver, where it can be phosphorylated to glycerol-3-phosphate. The phosphorylated glycerol can enter glycolysis or gluconeogenesis at the level of triose phosphates. If gluconeogenesis occurs, for every 2 moles of glycerol-3-phosphate, 1 mole of glucose can be synthesized.

297. The answer is D. *(Stryer, 4/e, pp 605–606.)* Fats (triacylglycerols) are the most highly concentrated and efficient stores of metabolic energy in the body. This is because they are anhydrous and reduced. On a dry-weight basis, the yield from the complete oxidation of the fatty acids produced from triacylglycerols is approximately 9 kcal/g compared with 4 kcal/g for glycogen and proteins. However, under physiologic conditions, glycogen and proteins become highly hydrated, whereas triacylglyceride stores remain relatively free of water. Therefore, while the energy yield from fat stores remains at approximately 9 kcal/g, the actual yields from the oxidation of glycogen and proteins are diluted considerably. Under anhydrous physiologic conditions, fats yield about six times the energy of glycogen stores.

298. The answer is D. *(Stryer, 4/e, pp 75–77.)* The diagram shown is actually of ADP, which is a nucleotide. However, since the sugar in the ADP is ribose, it is most correctly and completely described as a ribonucleotide. If the sugar was deoxyribose and lacked one of the two hydroxyl groups shown, then it would have been described as a deoxyribonucleotide. Recall that a nucleoside is a nitrogenous purine or pyrimidine base attached to a pentose sugar. The pentose sugar may either be a ribose or deoxyribose. A nucleotide is a phosphate ester of a nucleoside. The phosphate ester may be formed of one, two, or three phosphate groups (e.g., AMP, ADP, or ATP).

299. The answer is E. *(Stryer, 4/e, pp 559–565.)* The pentose phosphate pathway generates reducing power in the form of NADPH in the oxidative branch of the pathway and synthesizes 5-carbon sugars in the nonoxidative branch of the pathway. The pentose phosphate pathway also carries out the interconversion of 3-, 4-, 5-, 6-, and 7-carbon sugars in the nonoxidative reactions. The final sugar product of the oxidative branch of the pathway is ribulose-5-phosphate. The first step of the nonoxidative branch of the pathway is the conversion of ribulose-5-phosphate to ribose-5-phosphate or xylulose-5-phosphate in the presence of the enzymes phosphopentose isomerase and phosphopentose epimerase, respectively. Thus ribulose-5-phosphate is a key intermediate that is common to both the oxidative and nonoxidative branches of the pentose phosphate pathway.

300. The answer is D. *(Stryer, 4/e, pp 694–695.)* In the first stage of cholesterol formation, acetyl coenzyme A condenses to form mevalonate, which is then phosphorylated and decarboxylated to form isopentenyl pyrophosphate. Half of the isopentenyl pyrophosphate isomerizes to form dimethylallyl pyrophosphate. These two isomeric C_5 pyrophosphate units (isopentenyl pyrophosphate and dimethylallyl pyrophosphate) condense to form a C_{10} compound called *geranyl pyrophosphate.* Isopentenyl pyrophosphate then condenses with geranyl pyrophosphate to form the C_{15} compound farnesyl pyrophosphate. Finally, two farnesyl pyrophosphates condense in the presence of NADPH to form the C_{30} compound squalene. Squalene is ultimately cyclized through a series of steps to form cholesterol. Thus, the correct sequence of events leading from C_5 units to C_{30} squalene is sequential condensation of 5-carbon units until a 15-carbon unit is formed, then condensation of two 15-carbon units to form squalene.

301. The answer is A. *(Stryer, 4/e, pp 567–568.)* One of the world's most common enzyme deficiencies is glucose-6-phosphate-dehydrogenase deficiency. This deficiency in erythrocytes is particularly prevalent among African and Mediterranean peoples. A deficiency in glucose-6-phosphate dehydrogenase blocks the pentose phosphate pathway and NADPH production. Without NADPH to maintain glutathione in its reduced form, erythrocytes have no protection from oxidizing agents. This inherited deficiency is often diagnosed when patients develop hemolytic anemia after receiving oxidizing drugs such as pamaquine or after eating oxidizing substances such as fava beans.

302. The answer is C. *(Stryer, 4/e, pp 511–514.)* In the citric acid cycle, the conversion of α-ketoglutarate to succinate results in decarboxylation, transfer of an H^+/e^- pair to NADH + H^+, and the substrate-level phosphorylation of GDP to GTP. The series of reactions involved is quite complex. First, α-ketoglutarate reacts with NAD^+ + CoA to yield succinyl CoA + CO_2 + NADH + H^+. These reactions occur by the catalysis of the α-ketoglutarate dehydrogenase complex, which contains lipoamide, FAD^+, and thiamine pyrophosphate as prosthetic groups. Under the action of succinyl CoA synthetase, succinyl CoA catalyzes the phosphorylation of GDP with inorganic phosphate coupled to the cleavage of the thioester bond of succinyl CoA. Thus, the production of succinate from α-ketoglutarate yields one substrate-level phosphorylation and the production of 3 ATP equivalents from NADH via oxidative phosphorylation.

303. The answer is E. *(Stryer, 4/e, pp 622–623.)* In mammals, a variety of fatty acids are considered essential and cannot be synthesized. These include linoleate (C-18 *cis*-Δ^9,Δ^{12}) and linolenate (C-18 *cis*-$\Delta^9,\Delta^{12},\Delta^{15}$). Either these fatty acids or fatty acids to which they are precursors must be supplied in the diet as starting points for synthesis of a variety of other unsaturated fatty acids that will lead to the synthesis of prostaglandins, thromboxanes, and leukotrienes. For example, arachidonate, a 20-carbon fatty acid with four double bonds is derived from linolenate. Arachidonate gives rise to some prostaglandins, thromboxanes, and leukotrienes. Some fatty acids must be obtained in the diet because of the limitations governing enzymes of fatty acid synthesis in humans; that is, double bonds cannot be introduced beyond the 9–10 bond position of carbons in the fatty acid chain, and subsequent double bonds after the first must be separated by two single bonds. Thus, linolenate and linoleate cannot be synthesized in humans.

304. The answer is B. *(Stryer, 4/e, pp 495–496.)* The conversion of glucose to glucose-6-phosphate is different in liver and muscle. In muscle and most other tissues, hexokinase regulates the conversion of glucose to glucose-6-phosphate. When the major regulatory enzyme of glycolysis, phosphofructose kinase, is turned off, the level of fructose-6-phosphate increases and in turn the level of glucose-6-phosphate rises because it is in equilibrium with fructose-6-phosphate. Hexokinase is inhibited by glucose-6-phosphate. However, in liver, glucose is phosphorylated even when glucose-6-phosphate levels are high because the enzyme regulating glucose transformation into glucose-6-phosphate is glucokinase. Glucokinase is not inhibited by glucose-6-phosphate in liver. While hexokinase has a low K_m for glucose and is capable of acting upon low levels of blood glucose, glucokinase has a high K_m for glucose and is effective only when glucose is abundant. Therefore, when blood glucose levels are low, muscle, brain, and other tissues are capable of taking up and phosphorylating glucose, while liver is not. When blood glucose is abundant, glucokinase in liver phosphorylates glucose and provides glucose-6-phosphate for the synthesis and storage of glucose as glycogen.

305. The answer is C. *(Stryer, 4/e, pp 488–491.)* All the enzymes named are glycolytic enzymes that carry out phosphorylation of glucose-derived substrates or of ADP to form ATP. However, only the reaction catalyzed by glyceraldehyde-3-phosphate dehydrogenase is a phosphorylation reaction coupled to oxidation that uses inorganic phosphate. In this reaction, glyceraldehyde-3-phosphate is converted to 1,3-bisphosphoglycerate by the addition of inorganic phosphate and the oxidation of glyceraldehyde-3-phosphate with the concomitant reduction of NAD^+ to NADH + H^+. This reaction is an example of a high-energy phosphate compound being produced by an oxidation-reduction reaction. The oxidation of the aldehyde group at C-1 of glyceraldehyde-3-phosphate provides the energy for the reaction. The 1,3-bisphosphoglycerate can then be utilized to phosporylate ADP to ATP through the action of phosphoglycerate kinase, which is the next step in the glycolytic pathway.

306. The answer is C. *(Stryer, 4/e, pp 594–596.)* Muscle contraction is caused by the release of calcium from the sarcoplasmic reticulum following nervous stimulation. In addition to stimulating contraction, the calcium released from the sarcoplasmic reticulum binds to a calmodulin subunit on phosphorylase kinase. This activates phosphorylase kinase, converting it from the D form to the A form. The activated phosphorylase then breaks down glycogen and provides glucose for energy metabolism during exercise. In this way, muscle contraction and glucose production from glycogen are coordinated by the transient increase of cytoplasmic calcium levels during muscle contraction.

307. The answer is B. *(Stryer, 4/e, pp 685–689.)* Diacylglycerol-3-phosphate, more commonly known as *phosphatidate,* is an intermediate common to the synthesis of both triacylglycerol and phospholipids. In a two-step process, glycerol phosphate is successively acylated by two acyl CoA's to lysophosphatidate, which contains a fatty acid group in the 1′ position, and then phosphatidate, which contains fatty acid groups in the 1′ and 2′ position with a phosphate group in the 3′ position. From that point, pathways for synthesis of phospholipids and triacylglycerol diverge. If storage lipid is to be produced, phosphatidate is dephosphorylated by a phosphatase and then acylated by acyl CoA to form triacylglycerol. In contrast, if phospholipids are to be produced, phosphatidate is activated by CTP in a reaction that produces CDP-diacylglycerol and pyrophosphate. Phosphatidylserine, phosphatidylinositol, phosphatidylethanolamine, and phosphatidylcholine can all be derived from CDP-diacylglycerol.

308. The answer is D. *(Stryer, 4/e, pp 614–621.)* Two major enzyme complexes are involved in the synthesis of fatty acids. The first is acetyl CoA carboxylase, which synthesizes malonyl CoA by the steps shown below for the synthesis of palmitate:

$$7 \text{ Acetyl CoA} + 7 \text{ HCO}_3^- + 7 \text{ ATP} \rightarrow 7 \text{ malonyl CoA} + 7 \text{ ADP} + 7 \text{ P}_i$$

Using the malonyl CoA, palmitate is then synthesized by seven cycles of the fatty acid synthetase complex whose stoichiometry is summarized below:

$$\text{Acetyl CoA} + 7 \text{ malonyl CoA} + 14 \text{ NADPH} \rightarrow \text{palmitate} + 7 \text{ CO}_2 + 14 \text{ NAD}^+ + 8 \text{ CoA} + 6 \text{ H}_2\text{O}$$

As can be seen from the equations above, the necessary amount of malonyl CoA is synthesized. Palmitate is subsequently synthesized from malonyl CoA and one initial acetyl CoA. Thus, acetyl CoA, NADPH, ATP, and HCO_3^- are all necessary in this process. In contrast, FADH is not utilized in fatty acid synthesis, but is one of the products of fatty acid oxidation.

309. The answer is D. *(Stryer, 4/e, pp 689–691.)* Both sphingomyelin and gangliosides are end products derived from ceramides, the structural units of sphingolipids. Both sphingomyelin and gangliosides are sphingolipids. Sphingomyelin is derived by the contribution of phosphorylcholine from CDP-choline to the terminal hydroxyl group of ceramide. In contrast, in order to form gangliosides, the terminal phosphate group of ceramide is substituted with glucose or galactose from UDP-glucose or UDP-galactose. This forms a cerebroside. An oligosaccharide is linked to the glucose residue of the ceramide. Thus, the structure of a ganglioside is equivalent to a glycolipid, while the structure of sphingomyelin is quite similar to that of the phospholipid phosphatidylcholine. Palmitoyl CoA and serine react and through a series of steps form sphingosine, the immediate precursor of ceramide. Thus, palmitoyl CoA, serine, sphingosine, and cerebrosides can be considered precursors of gangliosides, while sphingomyelin cannot.

310. The answer is B. *(Stryer, 4/e, pp 474–478, 920–921.)* Glycoproteins are formed by the modification of preexisting proteins in the lumen of the endoplasmic reticulum. During this modification, oligosaccharide units are attached to the side-chain oxygen atoms of serine or threonine residues by *O*-glycosidic linkages. Alternatively, oligosaccharide units are attached to the side-chain nitrogen of asparagine residues by *N*-glycosidic linkages. During glycosylation of proteins in the endoplasmic reticulum, a large, common oligosaccharide is transferred to the polypeptide chain after being attached to the carrier dolichol phosphate. Dolichol phosphate is a long lipid composed of repeating isoprene units. The site of attachment of the activated oligosaccharide is the terminal phosphoryl group of this lipid. Once the core oligosaccharide is developed on the dolichol phosphate, it is transferred to the protein. Any integral membrane proteins are glycoproteins with oligosaccharides on the external surface. In addition, secreted proteins, including clotting factors and antibodies, are glycoproteins.

311. The answer is A. *(Stryer, 4/e, pp 471–472.)* Most of fructose found in the diet of North Americans is derived from the disaccharide sucrose. Sucrose is common table sugar. Sucrose is cleaved into equimolar amounts of glucose and fructose in the small intestine by the action of the pancreatic enzyme sucrase.

312. The answer is E. *(Stryer, 4/e, pp 313–314.)* Treatment of patients with congestive heart failure is often based upon the use of cardiotonic steroids such as digitalis. Digitalis is derived from the foxglove plant and has been used as an herbal cure for heart problems since ancient times. Digitalis and ouabain are cardiotonic steroids that inhibit the Na^+, K^+ -ATPase pump located in the plasma membrane of cardiac muscle cells. They specifically inhibit the dephosphorylation reaction of the ATPase when the cardiotonic steroid is bound to the extracellular face of the membrane. By inhibiting the pump, higher levels of sodium are left inside the cell, leading to a diminished sodium gradient. This results in a slower exchange of calcium by the sodium-calcium exchanger. Subsequently, intracellular levels of calcium are maintained at a higher level and greatly enhance the force of contraction of cardiac muscle.

313. The answer is B. *(Stryer, 4/e, pp 471–473.)* Lactose is a disaccharide of glucose and galactose found in milk. Amylose is an unbranched polymer of glucose residues in α-1,4 linkages. Glycogen is a branched polymer of glucose with both α-1,4 and α-1,6 linkages. Maltose is a disaccharide of glucose, which is usually the breakdown product of amylose. Cellulose, the most abundant compound known, is the structural fiber of plants and bacterial walls. It is an unbranched polymer of glucose residues in α-1,4 linkages in a β-configuration.

314. The answer is A. *(Stryer, 4/e, pp 593–595.)* The sequential cleavage of the α-1,4-glycosidic bonds of glycogen to release successive glucose-1-phosphate residues is known as *glycogenolysis.* The enzyme catalyzing this reaction is glycogen phosphorylase *a,* an active, phosphorylated tetramer formed by covalent modification of phosphorylase *b,* an inactive dimer. In glycogenesis or glycogen synthesis, activated glycogen synthase adds the glucose of uridine diphosphate (UDP)–glucose units to a growing glycogen polymer by forming α-1,4 linkages. In contrast to phosphorylase, glycogen synthase is inactivated by covalent phosphate binding. The same enzyme that inactivates glycogen synthase by catalyzing its phosphorylation activates another enzyme, phosphorylase kinase, which activates glycogen phosphorylase by phosphorylation. The hormone-stimulated cascade that stimulates glycogenolysis and turns off glycogenesis is shown in the accompanying figure.

Second-Messenger Cascade of Activation of Glycogenolysis and Inactivation of Glycogenesis

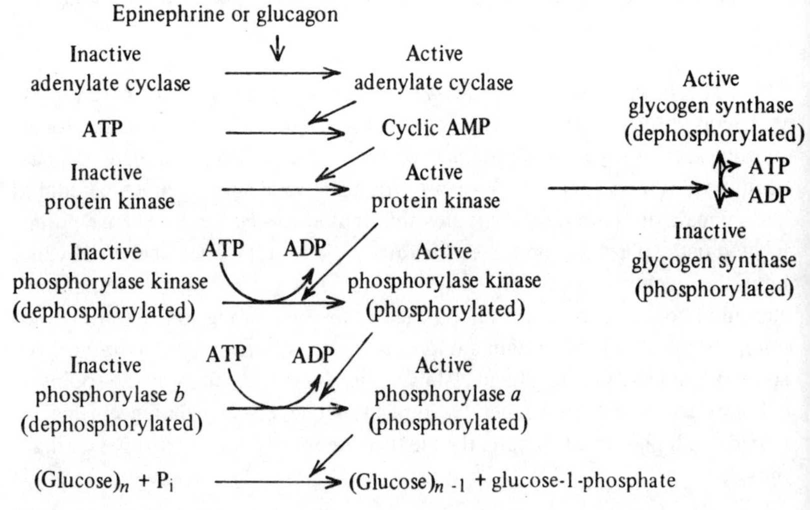

315. The answer is C. *(Stryer, 4/e, pp 570–572, 594–595, 614–615.)* During an overnight fast, glycogenolysis and gluconeogenesis occur to some degree. Following a well-rounded breakfast, amino acids are available for protein synthesis, glycogen synthesis occurs from excess glucose, gluconeogenesis decreases, and fatty acid synthesis occurs from excess acetyl CoA produced from dietary sources. Thus, the activity of enzymes of gluconeogenesis (pyruvate carboxylase and phosphoenolpyruvate carboxykinase) decreases, while the activity of enzymes of fatty acid synthesis (acetyl CoA carboxylase) increases.

316. The answer is A. *(Stryer, 4/e, pp 705–706.)* Cortisol, the major glucocorticoid, and aldosterone, the major mineralocorticoid, as well as the sex hormones testosterone, estrone, and estradiol are all synthesized from a common precursor—progesterone. The first steps in the synthesis of steroid hormones are the hydroxylation and cleavage of the 6-carbon side chain of cholesterol (C_{27}) to yield pregnenolone (C_{21}). The actual cleavage of the side chain is catalyzed by the enzyme desmolase. The 3-hydroxyl of pregnenolone is oxidized to a 3-keto group and the Δ^5 double bond isomerized to a Δ^4 double bond to form progesterone.

317. The answer is A. *(Stryer, 4/e, pp 689–691.)* Gangliosides appear to serve special functions in the cell plasma membrane as components of receptors for viruses, peptide hormones, and pharmacologically active substances. They are composed of sphingosine, a long-chain fatty acid, hexose(s), and neuraminic acid or its *N*-acetyl derivative, sialic acid. Gangliosides do not contain glycerol, which forms the backbone of glycerides and phosphatides.

318. The answer is D. *(Stryer, 4/e, pp 622–624.)* Many of the eicosanoids (20-carbon compounds)—prostaglandins, thromboxanes and leukotrienes—are derived from arachidonic acid. The scientific name of arachidonic acid is *eicosatetraenoic acid.* Arachidonic acid can only be synthesized from essential fatty acids obtained from the diet. The essential fatty acid linoleic acid ($C_{18}\Delta^{9,12}$) is desaturated to form γ-linolenic acid ($C_{18}\Delta^{6,9,12}$), which is sequentially elongated and desaturated to form eicosatrienoic acid ($C_{20}\Delta^{8,11,14}$) and arachidonic acid ($C_{20}\Delta^{5,8,11,14}$), respectively. Eicosanoic acid (C_{20}) is arachidic acid, a nonessential fatty acid.

319. The answer is D. *(Stryer, 4/e, pp 622–624.)* Prostaglandins are derived from unsaturated fatty acids containing 20 carbons; for example, PGE_2 is derived from arachidonic acid. They contain at least three double bonds, one or more hydroxyl groups, and oxygen groups. A cyclopentane ring is also present. The carboxyl end of the original fatty acid is retained. No methyl branches are present.

320. The answer is A. *(Stryer, 4/e, pp 687–690.)* The backbone of all sphingolipids is sphingosine, a long-chain aliphatic amine that is formed from palmitate and serine. The addition of another fatty acid to the amino group yields ceramide (*N*-acylsphingosine). Ceramide serves as the starting point for the synthesis of all sphingolipids, where the terminal hydroxyl group is substituted. In cerebrosides, glucose or galactose is linked to the terminal hydroxyl of ceramide. In gangliosides, an oligosaccharide can be linked to the glucose residue. Phosphocholine is attached to the terminal hydroxyl of ceramide to form sphingomyelin. Except for the sphingosine backbone, sphingomyelin is quite similar to lecithin (phosphatidyl choline), which is a phospholipid with a glycerol backbone.

321. The answer is E. *(Stryer, 4/e, pp 693, 708–709.)* Isopentenyl pyrophosphate and dimethylallyl pyrophosphate are isoprenoid isomers formed from the repeated condensation of acetyl CoA units. By continued condensation in mammalian systems, cholesterol can be formed. In plant systems carotenoids are formed. In addition to producing the color of tomatoes and carrots, they serve as the light-absorbing molecules of photosynthesis. In mammals, β-carotene is the precursor of retinal, which is the basic chromophore of all visual pigments. Ketone bodies are also derived from condensation of acetyl CoA units but are not derived from isoprenoid units.

322. The answer is E. *(Stryer, 4/e, pp 593–595.)* Activated phosphorylase kinase activates glycogen phosphorylase, which catalyzes the breakdown of glycogen. Phosphorylase kinase can be activated in two ways. Phosphorylation leads to complete activation of phosphorylase kinase. In the presence of low blood glucose, epinephrine or norepinephrine interacts with specific receptors to stimulate adenylate cyclase production of cyclic AMP. Cyclic AMP activates protein kinase, which catalyzes phosphorylation and activation of phosphorylase kinase. Alternatively, in muscle, the transient increases in levels of Ca^{2+} associated with contraction lead to a partial activation of phosphorylase kinase. Ca^{2+} binds to calmodulin, which is a subunit of phosphorylase kinase. Calmodulin regulates many enzymes in mammalian cells through Ca^{2+} binding.

323. The answer is D. *(Stryer, 4/e, pp 577–578.)* In mammalian systems ATP may be formed in the presence or the absence of molecular oxygen. In glycolysis, ATP is formed by substrate-level phosphorylation. In mitochondria, ATP is formed by the oxidation of NADH. ATP hydrolysis is an exergonic reaction that releases energy. When this released energy is chemically linked to an energy-requiring or endergonic reaction, that reaction is driven. Heat does not drive biologic reactions.

324. The answer is A. *(Stryer, 4/e, pp 608–610.)* During each round of ongoing β-oxidation, saturated fatty acyl CoA is degraded by being shortened by 2 carbons released in the form of acetyl CoA. In all, four segmental reactions take place: oxidation of acyl CoA to enoyl CoA to generate $FADH_2$ from FAD^+; hydration with the consumption of H_2O to yield L-hydroxyacyl CoA; oxidation to ketoacyl CoA linked to production of NADH + H^+ from NAD^+; and cleavage of ketoacyl CoA by addition of new CoA to yield fatty acyl CoA and acetyl CoA.

325. The answer is A. *(Stryer, 4/e, pp 585–586.)* The hydrolysis of ATP to AMP plus pyrophosphate (PP_i) is reversible. The hydrolysis of the pyrophosphate (PP_i) bond is –4.5 kcal/mole. Thus, for all intents and purposes, the coupling of pyrophosphate to AMP does not occur. Instead, phosphorylation of AMP by ATP to yield 2 ADP is the first step in the reformation of ATP from AMP. It is catalyzed by adenylate kinase. ADP then reacts with inorganic phosphate (P_i) to form ATP during oxidative phosphorylation. CDP, GDP, TDP, and UDP can all be phosphorylated to the triphosphate form by specific nucleoside diphosphokinases.

326. The answer is B. *(Stryer, 4/e, pp 539–541, 703–704.)* Some monooxygenases found in liver endoplasmic reticulum require cytochrome P_{450}. This cytochrome acts to transfer electrons between NADPH, O_2, and the substrate. It can be an electron acceptor from a flavoprotein. In the mitochondrial electron transport chain, flavoproteins donate electrons to coenzyme Q, which then transfers them to other cytochromes. Flavoproteins that are oxidases often react directly with molecular oxygen to form hydrogen peroxide. Flavoproteins can be NADH dehydrogenases that oxidize NADH and transfer the electrons to coenzyme Q. The electron transfer centers of flavoproteins in the electron transport chain contain nonheme iron and sulfur.

327. The answer is C. *(Stryer, 4/e, pp 570–574.)* In the formation of phosphoenolpyruvate during gluconeogenesis, oxaloacetate is an intermediate. In the first step, catalyzed by pyruvate carboxylase, pyruvate is carboxylated with the utilization of one high-energy ATP phosphate bond.

$$\text{Pyruvate} + \text{ATP} + CO_2 \longleftrightarrow \text{oxaloacetate} + \text{ADP} + P_i$$

In the second step, catalyzed by phosphoenolpyruvate carboxykinase, a high-energy phosphate bond of GTP drives the decarboxylation of oxaloacetate.

$$\text{Oxaloacetate} + \text{GTP} \longleftrightarrow \text{phosphoenolpyruvate} + \text{GDP} + CO_2$$

In contrast to gluconeogenesis, the formation of pyruvate from phosphoenolpyruvate during glycolysis requires only pyruvate kinase, and ATP is made.

328. The answer is A. *(Stryer, 4/e, pp 263–272.)* The classic chemical definition of a lipid is a substance that is poorly soluble in water but soluble in organic solvents. Lipids are hydrophobic, water-insoluble substances that may contain phosphate and nitrogen in addition to carbon, hydrogen, and oxygen. The main storage form of lipids is triacylglycerols, which are completely insoluble in water. The main structural form of lipids is phospholipids, which are partially soluble in water since they are amphipathic; that is, phospholipids possess a hydrophilic (water-loving) head group in addition to their hydrophobic (water-hating) hydrocarbon tails. Lipids associate in droplets, micelles, or bilayers by noncovalent, hydrophobic interactions.

329. The answer is C. *(Stryer, 4/e, pp 612–613.)* Starvation results in the increased use of lipids as an energy source, with increased oxidation of fatty acids and production of acetoacetyl CoA, a precursor of ketone bodies. Ketone bodies include acetoacetic acid and β-hydroxybutyrate, which are formed in the liver, and acetone, which is spontaneously formed from excess acetoacetate in the blood. Although the liver forms ketone bodies from excess acetyl CoA produced by β-oxidation, it cannot use ketone bodies. The utilization of ketone bodies for metabolism is also seen in insulin deficiency (diabetes mellitus). Diabetes mellitus, like starvation, results in increased mobilization of fatty acids and, hence, increased production of ketone bodies by the liver.

330. The answer is D. *(Stryer, 4/e, pp 613–616.)* Pyruvate and acetyl CoA are important intermediates in the conversion of glucose to fatty acids. The acetyl CoA exits from the mitochondrion in the form of citrate, which is converted to cytosolic acetyl CoA and oxaloacetate. Acetyl CoA is converted to malonyl CoA by acetyl CoA carboxylase, the regulating enzyme of fatty acid synthesis. From that point on all the substrates of the growing fatty acid chain are attached to the acyl carrier protein (ACP) portion of the fatty acid synthase multienzyme complex. Therefore, 3-ketoacyl ACP, not 3-ketoacyl CoA, is an intermediate of fatty acid synthesis. However, 3-ketoacyl CoA is an intermediate of fatty acid β-oxidation.

331. The answer is C. *(Stryer, 4/e, pp 584–588.)* Blood glucose is rapidly converted to glucose-6-phosphate upon entering cells by hexokinase or, in the case of liver, by glucokinase. Glucose-6-phosphate is in equilibrium with glucose-1-phosphate via the action of phosphoglucomutase. Glucose-1-phosphate is activated by UTP to form UDP-glucose, which is added to glycogen by an α-1,4 linkage in the presence of glycogen synthase. To increase the solubility of glycogen and to increase the number of terminal residues, glycogen-branching enzyme transfers a block of about seven residues from a chain at least eleven residues long to a branch point at least four residues from the last branch point. The branch is attached by an α-1,6 linkage.

332. The answer is C. *(Stryer, 4/e, pp 685–691.)* Phosphoglycerides contain glycerol, phosphate, two fatty acid molecules, and either choline, ethanolamine, serine, or inositol bound in an ester linkage to phosphoric acid. Depending on their constitution, they may be designated specifically *phos-*

phatidylcholine, phosphatidylethanolamine, phosphatidylserine, or *phosphoinositide.* In contrast, the backbone of sphingolipids is sphingosine, which is joined to a long-chain fatty acid to form ceramide. Ceramide may be reacted with choline to form sphingomyelin or reacted with sugars to form gangliosides, which may have complex oligosaccharide head groups.

333. The answer is C. *(Stryer, 4/e, pp 612–613, 693.)* 3-Hydroxy-3-methylglutaryl CoA is not an inhibitor of cholesterol synthesis. The compound 3-hydroxy-3-methylglutaryl CoA is formed by the condensation of acetoacetyl CoA and acetyl CoA in the synthetic pathways for both cholesterol and ketone bodies. However, the similar enzymes involved in each pathway are separated in space. 3-Hydroxy-3-methylglutaryl CoA produced in mitochondria is cleaved to yield the ketone body acetoacetate, whereas that produced in the cytosol is reduced to form mevalonic acid, which goes on to form cholesterol. The two series of reactions are also separated in time. Synthesis of cholesterol occurs when excess acetyl CoA produced from carbohydrates is available. During fasting, synthesis of cholesterol is inhibited. In contrast, synthesis of ketone bodies is most rapid during fasting, when acetyl CoA is produced by the β-oxidation of mobilized fatty acids.

334. The answer is E. *(Stryer, 4/e, pp 570–572.)* Oxidative phosphorylation is the process by which ATP is formed when reducing equivalents of NADH and $FADH_2$ are transferred to O_2 by electron carriers in mitochondria. Glucose-6-phosphate is not involved in this process. However, glucose-6-phosphate is a pivotal compound in many pathways. Immediately upon entering cells, blood glucose is phosphorylated to glucose-6-phosphate by hexokinase in most cells and glucokinase in the liver. Glucose may only leave a cell in the dephosphorylated form produced by glucose-6-phosphatase, which is only found in liver. Glucose-6-phosphate may be the starting point of glycolysis, glycogen synthesis, and the pentose phosphate pathway. It can be considered the end point or switching point of glycogenolysis and gluconeogenesis.

335. The answer is B. *(Stryer, 4/e, pp 685–686.)* The activated intermediates in the synthesis of triacylglycerols are fatty acyl CoAs, which are added successively to glycerol-3-phosphate to produce lysophosphatidic acid; to lysophosphatidic acid to form phosphatidic acid, which is then dephosphorylated to diacylglycerol; and finally to diacylglycerol to form triacylglycerol. The activated intermediate CDP-diacylglycerol is utilized in the synthesis of phospholipids but not in the de novo synthesis of triacylglycerols.

336. The answer is D. *(Stryer, 4/e, pp 696, 735–736.)* Bile salts (acids) and lecithin (phosphatidyl choline) emulsify dietary lipids in the intestinal lumen by forming mixed micelles. Glycocholate is the major bile salt, while taurocholate is also present. Deoxycholate is a modification of cholic acid carried out by intestinal bacteria. Most bile salts (acids) are actively reabsorbed and reutilized. Bile salts are synthesized from cholesterol in the liver. Bile pigments are the products of heme catabolism that are excreted via the bile.

337. The answer is B. *(Stryer, 4/e, pp 503, 509–514, 752–754.)* Arsenite is an inhibitor of lipoic acid–containing enzymes such as α-ketoglutarate dehydrogenase. Malonate is an inhibitor of succinate dehydrogenase, and fluoracetate can be converted to fluorocitrate, which is an inhibitor of aconitase. The citric acid cycle requires oxygen and cannot proceed anaerobically. In contrast, fluorouracil is a suicide inhibitor of thymidylate synthase and blocks deoxythymidylate synthesis.

338. The answer is A. *(Stryer, 4/e, pp 581–582.)* Glycogen is a highly branched polymer of α-D-glucose residues joined by α-1,4-glycosidic linkage. The branched chains occur about every 10 residues and are joined in α-1, 6-glycosidic linkages. Large amounts of glycogen are stored as 100- to 400-Å granules in the cytoplasm of liver and muscle cells. The enzymes responsible for making or breaking the α-1,4-glycosidic bonds are contained within the granules. Thus glycogen is a readily mobilized form of glucose.

339. The answer is D. *(Stryer, 4/e, pp 514–518.)* The α-ketoglutarate dehydrogenase complex employs the cofactors NAD^+, FAD, thiamine pyrophosphate, lipoic acid, and CoA. This multienzyme complex yields succinyl CoA in a reaction that is virtually unidirectional because of energetics favoring the product. It is similar to the pyruvate dehydrogenase complex.

340. The answer is B. *(Stryer, 4/e, pp 686–688.)* There are a number of routes for the synthesis of phosphatidyl choline (also known as *lecithin*). In the de novo pathway, the activated intermediate is CDP-diacylglycerol, which combines with free choline to form phosphatidyl choline and CMP. Alternatively, choline can be phosphorylated to phosphorylcholine and then activated by reaction with CTP to form CDP-choline. In this pathway, also called the *salvage pathway,* CDP-choline is the activated intermediate that reacts with diacylglycerol to form phosphatidyl choline. Finally, in bacteria, phosphatidyl choline may be formed by the modification of phosphatidyl serine. In this pathway, phosphatidyl serine is decarboxylated to form phosphatidyl ethanolamine. Subsequently, the ethanolamine moiety is methylated three times to form the choline head group of phosphatidyl choline. *S*-Adenosylmethionine is the methyl donor.

341. The answer is C. *(Stryer, 4/e, pp 471–474.)* Mannose is a hexose monosaccharide common to oligosaccharides. Its stereochemical structure is different from that of glucose. Sucrose is common table sugar, usually made from cane or beet. It is a disaccharide composed of fructose linked to glucose by an α-glycosidic linkage. Lactose is the disaccharide of milk. It is formed from galactose and glucose by an α-glycosidic linkage. Glycogen is the major storage form of glucose in mammals. It is a polymer of glucose units linked by α-glycosidic linkages. Maltose is simply the disaccharide hydrolysis product of glycogen or starch. Digestion of the disaccharides occurs by the action of maltase, lactase, and sucrose, which are found in the small intestine.

342. The answer is E. *(Stryer, 4/e, pp 493–496.)* The glycolytic pathway has three key irreversible enzymes: hexokinase, phosphofructose kinase, and pyruvate kinase. Under conditions of cellular "plenty," ATP and citrate accumulate. Both are negative effectors of phosphofructokinase. In contrast, at low-energy charge, ADP and AMP accumulate and positively regulate phosphofructokinase. When phosphofructokinase is inhibited, glucose-6-phosphate accumulates and shuts off hexokinase. ATP inhibits the regulatory enzymes of glycolysis, while a lower energy charge actually stimulates glycolysis.

343–344. The answers are 343-B, 344-E. *(Stryer, 4/e, p 465.)* The structures depicted are (A) ribose, (B) glucose, (C) mannose, (D) galactose, and (E) fructose. The first four monosaccharides shown are D-aldoses containing four carbons (ribose) or six carbons (glucose, mannose, and galactose). An aldose is so named if the carbonyl group is at the end of the chain, which indicates that the monosaccharide is an aldehyde derivative. When the carbonyl function is at any other position, the sugar is a ketone derivative or ketose. Fructose is a D-ketose. Many of the monosaccharides are quite soluble in water and insoluble in nonpolar solvents. Many are sweet and form white crystalline solids upon drying.

345–348. The answers are 345-C, 346-B, 347-A, 348-E. *(Stryer, 4/e, pp 596–599, 607–608.)* B-100 is the structural coat protein of very low-density lipoproteins (VLDLs) that marks their delipidated remnants, LDLs, for uptake into peripheral tissues. A genetic deficiency of the receptors that recognize and bind this protein to the surface of peripheral tissues leads to familial hypercholesterolemia. In this condition, levels of blood cholesterol are high in heterozygotes and very high in homozygotes. This genetic increase in blood cholesterol as well as dietary increases in blood cholesterol can be treated with mevalonin. Mevalonin, an analogue of mevalonic acid, acts as a feedback inhibitor of 3′-hydroxy-3′-methylglutaryl-CoA (MNG-CoA) reductase, the regulated enzyme of cholesterol synthesis. Effective treatment with mevalonin decreases levels of blood cholesterol.

Von Gierke's disease is characterized by an enlarged liver and severe hypoglycemia (low blood glucose), in addition to ketosis, hyperlipemia, and hyperuricemia. Glucose-6-phosphatase is deficient and found only in the liver and kidney.

CoA-activated long-chain fatty acids bind to carnitine before being translocated across mitochondrial inner membranes to the inside of the mitochondrial matrix. A lack of carnitine or a deficiency of translocase leads to a lack of β-oxidation of long-chain fatty acids in mitochondria. Long-chain fatty acids are the predominant form of fatty acids normally metabolized by humans. In cases where carnitine deficiencies exist, muscle cramping is observed following fasting, meals high in fat content, and exercise. Each of these conditions leads to high levels of fatty acids in muscle.

349–350. The answers are 349-B, 350-D. *(Stryer, 4/e, pp 603–605.)* The systematic names of unbranched fatty acids describe the chain length, whether or not the fatty acid contains double bonds, and the number of double bonds. The presence or absence of double bonds is denoted by the suffix *-anoate,* meaning no double bonds, or *-enoate,* which indicates the presence of at least one double bond. The lack of a prefix following the initial designation of chain length indicates the presence of only one double bond in a fatty acid described as *-enoate.* Thus, *octadec-anoate* has 18 carbons and no double bonds (stearate); *octadec-enoate* has 18 carbons and one double bond (oleate). *Eicos-anoate* has 20 carbons and no double bonds. In contrast, *eicosa-tetra-enoate* has 20 carbons and four double bonds. If the double bonds were in positions 5, 8, 11, and 14, eicosatetraenoate would be arachidonic acid.

351–354. The answers are 351-B, 352-B, 353-D, 354-G. *(Stryer, 4/e, pp 773–775.)* The balance and integration of the metabolism of fats and carbohydrates are mediated by the hormones insulin, glucagon, epinephrine, and norepinephrine. All of these hormones have acute effects upon metabolism. Insulin and glucagon are two antagonistic hormones that maintain the balance of sugar and fatty acids in blood. Insulin is produced by the beta cells of the pancreas and its release is stimulated by high levels of glucose in the blood. It has a number of effects, but its major effect is to allow the entry of glucose into cells. Insulin also allows the dephosphorylation of key regulatory enzymes. The consequence of these actions is to allow glycogen synthesis and storage in both muscle and liver, suppression of gluconeogenesis, acceleration of glycolysis, promotion of the synthesis of fatty acids, and promotion of the uptake and synthesis of amino acids into protein. All in all, insulin acts to promote anabolism. In contrast, when blood sugar levels get low, the alpha cells of the pancreas release glucagon. The main targets of glucagon are the liver and adipose tissue. In the liver, glucagon stimulates the cyclic AMP–mediated cascade that causes phosphorylation of phosphorylase and glycogen synthesis. This effectively turns off glycogen synthase and turns on glycogen phosphorylase, thereby causing a breakdown of glycogen and a production of glucose in liver, which ultimately will raise blood glucose levels. Glucagon stimulates gluconeogenesis and blocks glycolysis.

355–356. The answers are 355-A, 356-C. *(Stryer, 4/e, pp 569–572.)* The reaction pathway shown is that of gluconeogenesis. It is a process for the synthesis of glucose from noncarbohydrate precursors. Gluconeogenesis is carried out primarily in the liver and to a lesser extent in the kidney during times of glucose need. Three irreversible steps of glycolysis are bypassed in gluconeogenesis.

1. Phosphoenolpyruvate is synthesized from pyruvate in a two-step process. First, oxaloacetate is synthesized from pyruvate by the addition of CO_2 at the expense of ATP. This is catalyzed by pyruvate carboxylase.

$$CO_2 + \text{pyruvate} + H_2O + \text{ATP} \rightarrow \text{oxaloacetate} + 2\text{H} + \text{ADP} + P_i$$

Phosphoenolpyruvate carboxykinase then catalyzes the conversion of oxaloacetate.

$$\text{Oxaloacetate} + \text{GPT} \leftrightarrow \text{phosphoenolpyruvate} + CO_2 + \text{GDP}$$

2. Fructose-1,6-bisphosphate catalyzes the hydrolysis of phosphate to form fructose-6-phosphate.

$$\text{Fructose-1,6-bisphosphate} + H_2O \rightarrow \text{frucdose-6-phosphate} + P_i$$

3. Free glucose is formed in the presence of glucose-6-phosphatase.

$$\text{Glucose-6-phosphate} + H_2O \rightarrow \text{glucose} + P_i$$

While it is not an irreversible reaction, the formation of glyceraldehyde-3-phosphate requires reduction of 1,3-diphosphoglycerate in the presence of dehydrogenase.

$$\text{1,3-Bisphosphoglycerate} + \text{NADH} + H^+ \rightarrow$$
$$\text{glyceraldehyde-3-phosphate} + \text{NAD}^+ + P_i$$

357–360. The answers are 357-F, 358-D, 359-B, 360-C. *(Stryer, 4/e, pp 417–418, 510–514, 552–553.)* In contrast to glycolysis, the only site of substrate-level phosphorylation in the tricarboxylic acid cycle is the step catalyzed by succinyl-CoA synthetase. In this step, the cleavage of CoA from succinyl CoA to produce succinate is the utilization of a high-energy bond of CoA to phosphorylate GDP with organic phosphate to produce GTP. Since NADH generated by glycolysis in the cytoplasm cannot pass across the mitochondrial membrane, shuttles are used to bring the electrons into the mitochondria for oxidative phosphorylation. In the glycerol phosphate shuttle, NADH + H^+ in the cytoplasm reduces dihydroxyacetone phosphate to glycerol phosphate, which is capable of entering the mitochondria. In the mitochondria, the glycerol phosphate is oxidized back to dihydroxyacetone phosphate, which can then diffuse back out into the cytoplasm. During this process, flavin ($FADH_2$) is reduced and is capable of generating 2 ATP via oxidation by the respiratory chain. In contrast, the malate-aspartate shuttle allows the formation of 3 ATP equivalents for each mole of cytoplasmic NADH + H^+ generated. The malate-aspartate shuttle is found mainly in the heart and liver. The process of oxidative phosphorylation that is coupled to electron transport occurs because of the proton gradient maintained across the mitochondrial membrane. ATP is formed by mitochondrial ATPase by the movements of protons across this gradient. In the presence of substances like 2,4-dinitrophenol (DNP), oxidation of oxidative phosphorylation is uncoupled. This occurs because DNP carries the protons across the mitochondrial membrane, short circuiting the phosphorylations that normally occur. While this reaction is not biologically useful, it does mimic the normal uncoupling of phosphorylation that can occur under certain biologic conditions and is used to generate heat to maintain body temperature. This occurs in certain mammals adapted to cold, newborn mammals, and hibernating animals. In these animals, this process of thermogenesis occurs in specialized brown adipose tissue. The uncoupling protein is called *thermogenin.* Since electron transport is tightly coupled to phosphorylation, under physiologic conditions electrons will not flow through the electron transport chain to O_2 unless ADP is simultaneously phosphorylated to ATP. If the level of ADP is low, oxidative phosphorylation will not occur at as high a rate and the rate of oxygen consumption in tissue will decrease. Respiratory control is the regulation of the rate of oxidative phosphorylation by ADP levels.

Vitamins and Hormones

DIRECTIONS: Each question below contains four or five suggested responses. Select the **one best** response to each question.

361. Enzymatic amplification of epinephrine stimulation of lipolysis in adipocytes is thought to differ from epinephrine stimulation of glycogenolysis in the liver in which way?

(A) Glucagon, not epinephrine, stimulates lipolysis in adipocytes
(B) The mechanism of hormone-receptor interaction is thought to be fundamentally different in each tissue
(C) Phosphorylase kinase is activated directly by a second messenger in adipocytes, but not in the liver
(D) Adenosine 3′,5′-cyclic monophosphate (cyclic AMP) is the second messenger in adipocytes, but not in the liver
(E) Only protein kinase is interposed as an amplification factor between the second messenger and the physiologically important enzymes in adipocytes

362. Which of the following is noted in Cushing's syndrome, a tumor-associated disease of the adrenal cortex?

(A) Decreased production of epinephrine
(B) Excessive production of epinephrine
(C) Excessive production of vasopressin
(D) Excessive production of cortisol
(E) Decreased production of cortisol

363. Following a normal overnight fast and a cup of black coffee, a diabetic woman feels slightly nauseous and decides to skip breakfast. However, she does take her shot of insulin. This may result in

(A) heightened glycogenolysis
(B) hypoglycemia
(C) increased lipolysis
(D) glucosuria
(E) toxic shock syndrome

364. In normal humans, insulin secretion during constant intravenous administration of glucose is best characterized by which of the following curves? (Starting time of glucose administration = 0.)

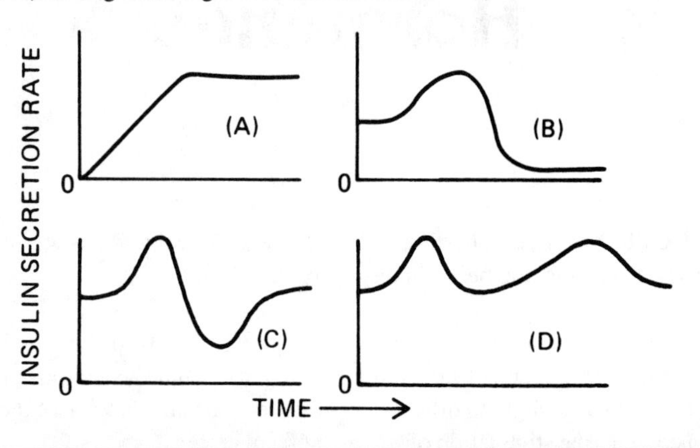

365. A cup of strong coffee would be expected to

(A) interfere with the synthesis of prostaglandins
(B) decrease the effects of glucagon
(C) enhance the effects of epinephrine
(D) provide the vitamin nicotinic acid
(E) do none of the above

366. Increased reabsorption of water from the kidney is the major consequence of which of the following hormones?

(A) Cortisol
(B) Insulin
(C) Vasopressin
(D) Glucagon
(E) Aldosterone

367. The important reactive group of glutathione in its role as an antioxidant is

(A) serine
(B) sulfhydryl
(C) tyrosine
(D) CoA
(E) carboxyl

368. A patient's glucose tolerance test is shown below. The most likely diagnosis is

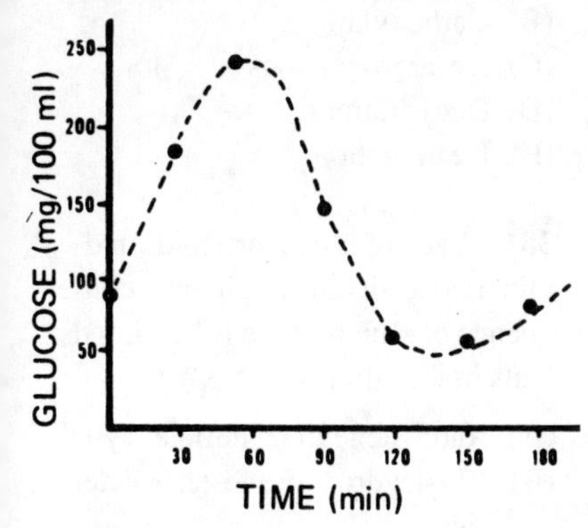

(A) severe diabetes
(B) Addison's disease
(C) von Gierke's disease
(D) delayed insulin response
(E) type I glycogen storage disease

369. Molecular iron, Fe, is

(A) stored primarily in the spleen
(B) stored in the body in combination with ferritin
(C) excreted in the urine as Fe^{2+}
(D) absorbed in the intestine by albumin
(E) absorbed in the ferric, Fe^{3+}, form

370. Humans most easily tolerate a lack of which of the following nutrients?

(A) Protein
(B) Iodine
(C) Carbohydrate
(D) Lipid
(E) Calcium

371. A deficiency of vitamin B_{12} causes

(A) cheliosis
(B) beriberi
(C) pernicious anemia
(D) scurvy
(E) rickets

372. In adults, a severe deficiency of vitamin D causes

(A) night blindness
(B) osteomalacia
(C) rickets
(D) skin cancer
(E) none of the above

373. Which of the following vitamins would most likely become deficient in a person who develops a completely carnivorous life-style?

(A) Thiamine
(B) Niacin
(C) Cobalamin
(D) Pantothenic acid
(E) Vitamin C

374. Which of the following statements regarding vitamin A is true?

(A) It is not an essential vitamin
(B) It is related to tocopherol
(C) It is a component of rhodopsin
(D) It is derived from ethanol
(E) It is also known as *opsin*

375. Pellagra can be prevented by treatment with

(A) thiamine
(B) niacin
(C) pyridoxine
(D) vitamin B_{12}
(E) pantothenic acid

376. Fully activated pyruvate carboxylase depends upon the presence of

(A) malate and niacin
(B) acetyl CoA and biotin
(C) acetyl CoA and thiamine pyrophosphate
(D) oxaloacetate and biotin
(E) oxaloacetate and niacin

377. Both Wernicke's disease and beriberi can be reversed by administering

(A) vitamin C
(B) vitamin A
(C) vitamin B_1
(D) vitamin B_6
(E) vitamin B_{12}

378. Pantothenic acid is a constituent of the coenzyme involved in

(A) decarboxylation
(B) acetylation
(C) dehydrogenation
(D) reduction
(E) oxidation

379. Studies of the action of two anticoagulants—dicumarol and warfarin (the latter also a hemorrhagic rat poison)—have revealed that

(A) vitamin C is necessary for the synthesis of fibrinogen
(B) vitamin C activates fibrinogen
(C) vitamin K is a clotting factor
(D) vitamin K is essential for γ-carboxylation of glutamate
(E) the action of vitamin E is antagonized by these compounds

380. Biotin is involved in which of the following types of reactions?

(A) Hydroxylations
(B) Carboxylations
(C) Decarboxylations
(D) Dehydrations
(E) Deaminations

381. Lack of glucocorticoids and mineralocorticoids might be a consequence of which of the following defects in the adrenal cortex?

(A) Androstenedione deficiency
(B) 17α-Hydroxyprogesterone deficiency
(C) Estrone deficiency
(D) C-21-Hydroxylase deficiency
(E) Testosterone deficiency

382. Which of the following vitamins is the precursor of CoA?

(A) Riboflavin
(B) Pantothenate
(C) Thiamine
(D) Cobamide
(E) Pyridoxamine

383. In the Far East, beriberi is a serious health problem. It is characterized by neurologic and cardiac symptoms. Beriberi is caused by a deficiency of

(A) choline
(B) ethanolamine
(C) thiamine
(D) serine
(E) glycine

384. Both acyl-carrier protein (ACP) of fatty acid synthetase and coenzyme A (CoA)

(A) contain reactive phosphorylated tyrosine groups
(B) contain thymidine
(C) contain phosphopantetheine reactive groups
(D) contain cystine reactive groups
(E) carry folate groups

385. Which one of the following transfers acyl groups?

(A) Thiamine pyrophosphate
(B) Lipoamide
(C) ATP
(D) NADH
(E) FADH

386. Which one of the following cofactors must be utilized during the conversion of acetyl CoA to malonyl CoA?

(A) Thiamine pyrophosphate
(B) Acyl-carrier protein (ACP)
(C) NAD^+
(D) Biotin
(E) FAD

387. ADP-ribosylation is the mode of action of

(A) cholera toxin
(B) acetyl choline
(C) muscarinic receptors
(D) cyclic AMP
(E) adenylate cyclase

388. Which one of the following hormones is derived most completely from tyrosine?

(A) Glucagon
(B) Thyroxine
(C) Insulin
(D) Prostaglandins
(E) Endorphins

389. Which one of the following enzymes requires a coenzyme derived from the vitamin whose structure is shown below?

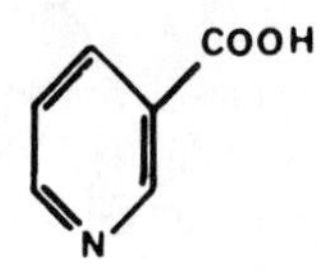

(A) Enoyl CoA hydratase
(B) Phosphofructokinase
(C) Glucose-6-phosphatase
(D) Glucose-6-phosphate dehydrogenase
(E) Glycogen phosphorylase

390. Cholera toxin causes massive and often fatal diarrhea by

(A) inactivating G_i protein
(B) irreversibly activating adenylate cyclase
(C) locking G_s protein into an inactive form
(D) rapidly hydrolyzing G protein GTP to GDP
(E) preventing GTP from interacting with G protein

391. Coenzymes derived from the vitamin shown below are required by enzymes involved in the synthesis of which of the following?

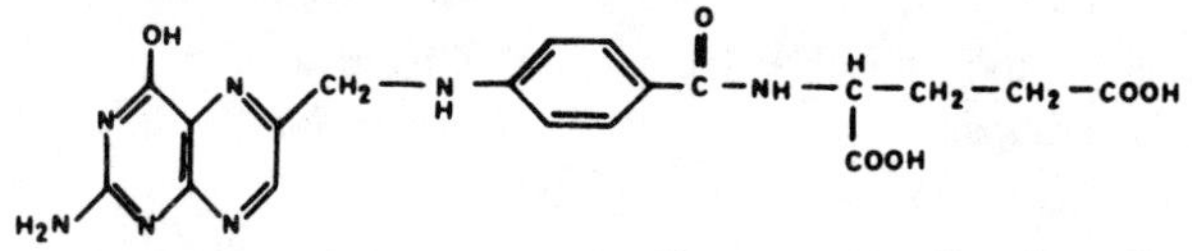

(A) ATP
(B) UTP
(C) CTP
(D) NADH
(E) NADPH

---

392. Coenzymes derived from the vitamin shown below are required by which one of the following enzymes?

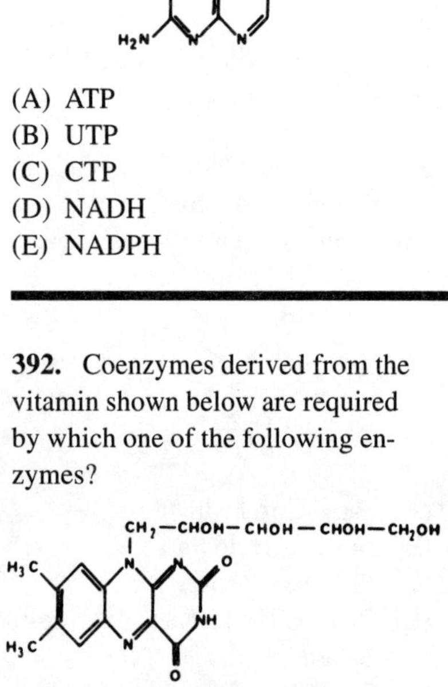

(A) Lactate dehydrogenase
(B) Glutamate dehydrogenase
(C) Pyruvate dehydrogenase
(D) Malate dehydrogenase
(E) Glyceraldehyde-3-phosphate dehydrogenase

393. Insulin regulates fatty acid synthesis by

(A) activating phosphorylase
(B) dephosphorylating of acetyl CoA carboxylase
(C) inhibiting malonyl CoA formation
(D) controlling carnitine-acyl CoA transferase activity
(E) activating fatty acid synthetase

394. Hormonal stimulation of the formation of the second messenger inositol 1,4,5-triphosphate (IP_3) quickly leads to the release of which other intracellular messenger?

(A) Cyclic AMP
(B) Prostaglandin
(C) Calcium
(D) Leukotriene
(E) Thromboxane

Questions 395–396

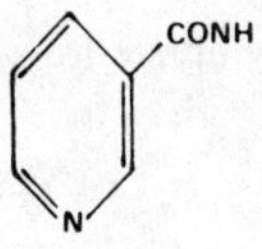

395. Which of the structures in the figure above is involved in amino transferase (transamination and deamination) and decarboxylation reactions of amino acids?

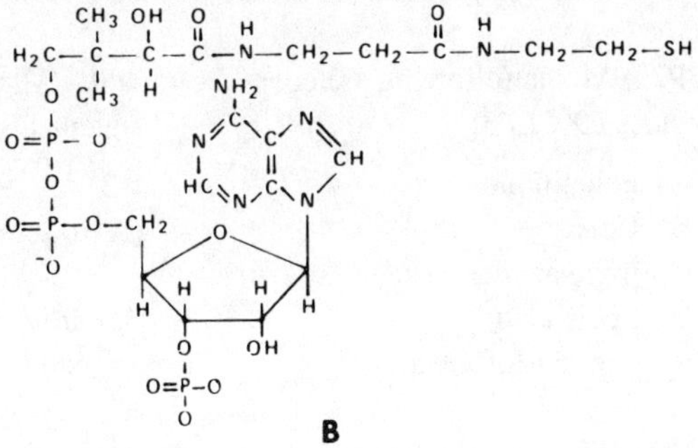

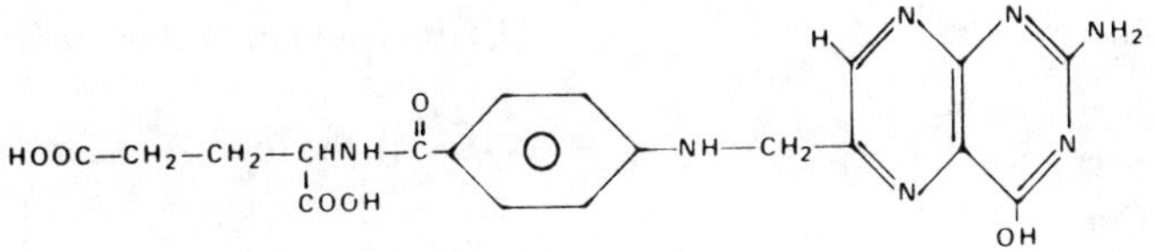

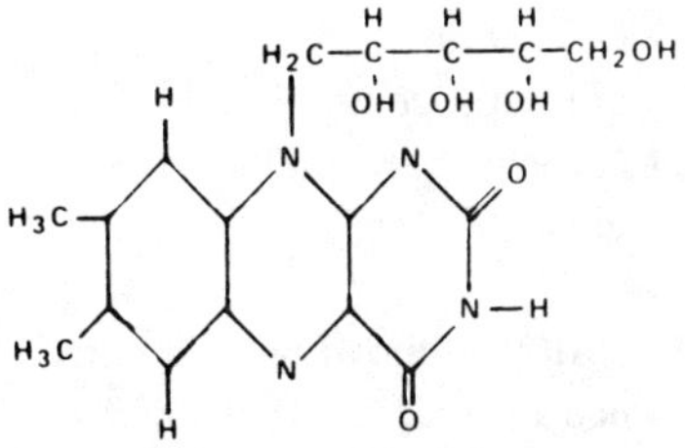

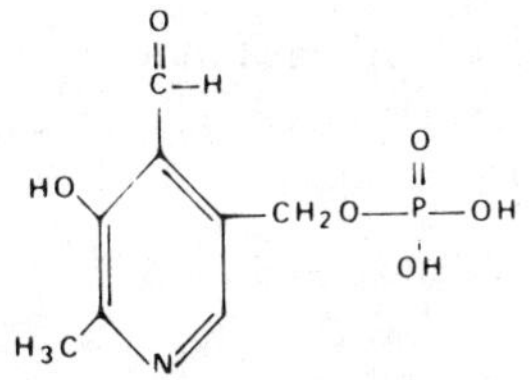

(A) A
(B) B
(C) C
(D) D
(E) E

396. Which of the compounds in the figure above is the precursor of the electron donor used in reductive biosynthesis?

(A) A
(B) B
(C) C
(D) D
(E) E

DIRECTIONS: Each numbered question or incomplete statement below is NEGATIVELY phrased. Select the **one best** lettered response.

397. All the following are coenzymes EXCEPT

(A) ubiquinone
(B) CoA
(C) pyruvate dehydrogenase
(D) lipoic acid
(E) tetrahydrofolic acid

398. Which one of the following is NOT a cofactor?

(A) Magnesium
(B) Iron
(C) Copper
(D) Methylcobalamin
(E) Zinc

399. All the following statements describing vitamin K are true EXCEPT that

(A) it is synthesized by intestinal bacteria
(B) it is obtained by eating spinach and cabbage
(C) it is obtained by eating egg yolk and liver
(D) it is required for liver synthesis of prothrombin
(E) it prevents thrombosis

400. All the following are true statements regarding vitamin A EXCEPT

(A) it promotes maintenance of epithelial tissue
(B) it promotes maintenance of vision
(C) it is synthesized in skin
(D) it is used to treat severe acne
(E) it is used to treat psoriasis

401. All the following hormones can cause a hyperglycemic effect EXCEPT

(A) thyroxine
(B) epinephrine
(C) glucocorticoids
(D) epidermal growth factor
(E) glucagon

402. All the following conditions produce a real or functional deficiency of vitamin K EXCEPT

(A) therapy with a coumarin derivative to prevent thrombosis in patients prone to clot formation
(B) prolonged, oral, broad-spectrum antibiotic therapy
(C) the total lack of red meat in the diet
(D) the total lack of green, leafy vegetables in the diet
(E) being a newborn infant

403. All the following statements correctly describe insulin EXCEPT

(A) it is thought to be an anabolic signal to cells that glucose is abundant
(B) it is converted from proinsulin to insulin primarily following secretion from β-cells
(C) it is inactive when in the proinsulin form
(D) it is a small polypeptide composed of two chains connected by disulfide bridges
(E) its action is antagonistic to that of glucagon

404. All the following hormones use cyclic AMP as a second messenger EXCEPT

(A) follicle-stimulating hormone
(B) luteinizing hormone
(C) glucagon
(D) estrogen
(E) epinephrine

405. Insulin has many direct effects on various cell types from such tissues as muscle, fat, liver, and skin. All the following cellular activities are increased following exposure to physiologic concentrations of insulin EXCEPT

(A) plasma membrane transfer of glucose
(B) glucose oxidation
(C) gluconeogenesis
(D) lipogenesis
(E) formation of ATP, DNA, and RNA

406. Following release of norepinephrine by sympathetic nerves and epinephrine by the adrenal medulla, all the following metabolic processes are increased EXCEPT

(A) glycolysis
(B) lipolysis
(C) gluconeogenesis
(D) glycogenolysis
(E) ketogenesis

407. All the following statements about prostaglandins are true EXCEPT

(A) they are cyclic fatty acids
(B) they have potent biologic effects that involve almost every organ in the body
(C) they were first observed to cause uterine contraction and lowering of blood pressure
(D) although found in many organs, they are synthesized only in the prostate and seminal vesicles
(E) medullin, isolated from the medulla of the kidney, is a prostaglandin

408. All the following compounds are members of the electron transport chain EXCEPT

(A) ubiquinone (coenzyme Q)
(B) cytochrome *c*
(C) nicotinamide adenine dinucleotide (NAD)
(D) flavin adenine dinucleotide (FAD)
(E) carnitine

409. The central ring structure shown below is common to all the following compounds EXCEPT

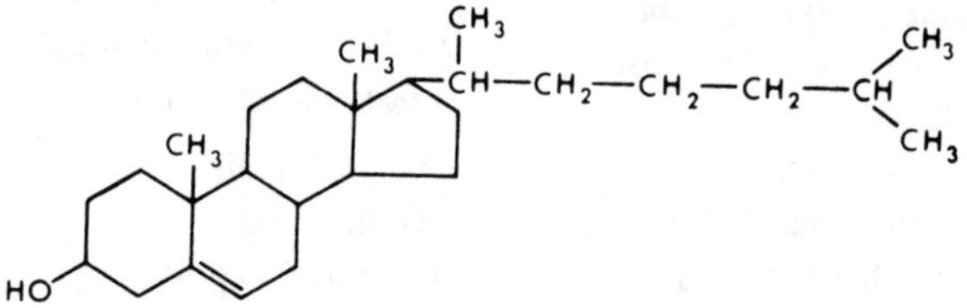

(A) adrenocorticotropin
(B) aldosterone
(C) testosterone
(D) bile acids
(E) vitamin D

---

410. All the following cofactors are involved in the pyruvate dehydrogenase reaction EXCEPT

(A) pyridoxal phosphate
(B) thiamine pyrophosphate
(C) lipoic acid
(D) FAD
(E) CoA

411. All the following statements regarding hormones are true EXCEPT that they

(A) may act as vitamins
(B) may influence the synthesis of enzymes
(C) may affect the rate of enzymatic catalysis
(D) may alter the permeability of plasma membranes
(E) act by regulating preexisting processes

412. Vitamins that function as dinucleotide derivatives include all the following EXCEPT

(A) vitamin B_2
(B) niacin
(C) nicotinate
(D) thiamine
(E) riboflavin

413. All the following vitamins give rise to cofactors that are phosphorylated in the active form EXCEPT

(A) pyridoxine
(B) lipoamide
(C) niacin
(D) thiamine
(E) riboflavin

414. Oxidative degradation of acetyl CoA in the citric acid cycle gives a net yield of all the following EXCEPT

(A) $FADH_2$
(B) 3 NADH
(C) 2 ATP
(D) 1 GTP
(E) 2 CO_2

415. All the following vitamins are soluble in organic solvents used to extract lipids from tissues EXCEPT

(A) vitamin A
(B) vitamin B_1
(C) vitamin D
(D) vitamin E
(E) vitamin K

416. Coenzyme A is common to all the following pathways EXCEPT

(A) citric acid cycle
(B) fatty acid synthesis
(C) fatty acid β-oxidation
(D) ketone body synthesis
(E) glycolysis

417. All the following hormones promote hyperglycemia EXCEPT

(A) epinephrine
(B) norepinephrine
(C) insulin
(D) glucagon
(E) glucocorticoids

418. In order to catalyze reactions, all the following enzymes require a phosphorylated derivative of the vitamin shown below EXCEPT

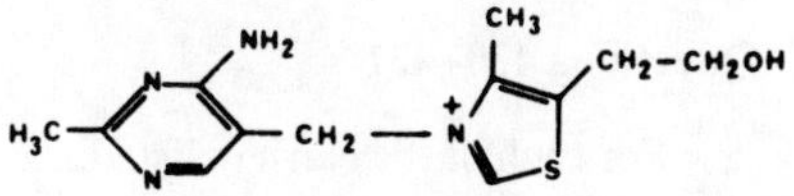

(A) pyruvate decarboxylase
(B) pyruvate dehydrogenase
(C) pentose phosphate shunt transketolase
(D) α-ketoglutarate dehydrogenase
(E) citrate synthetase

419. Coenzymes derived from the vitamin shown below are required by all the following enzymes EXCEPT

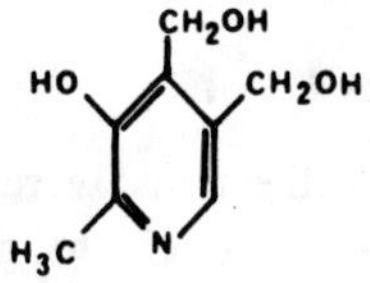

(A) alanine transaminase
(B) glutamate dehydrogenase
(C) glutamate transaminase
(D) glycogen phosphorylase
(E) tyrosine transaminase

DIRECTIONS: Each group of questions below consists of lettered headings followed by a set of numbered items. For each numbered item select the **one** lettered heading with which it is **most** closely associated. Each lettered heading may be used **once, more than once, or not at all.**

Questions 420–421

For each of the carriers listed, choose the activated group or units with which it is most closely associated.

(A) Electrons
(B) Phosphoryl
(C) Acyl
(D) Aldehyde
(E) Glucose

420. Uridine diphosphate (UDP)

421. NADPH

Questions 422–423

For each external secretion below, select the protein that governs it.

(A) Cholecystokinin
(B) Gastrin
(C) Insulin
(D) Intrinsic factor
(E) Secretin

422. Secretion of pancreatic juice into the intestine

423. Release of bile from the gallbladder

Questions 424–425

For each substance below, select the best description.

(A) A depositor of calcium in endoplasmic reticulum
(B) An inhibitor of protein kinase C
(C) A second messenger produced by the action of phospholipase C
(D) An ionophore
(E) A calcium detector

424. Inositol 1,4,5-triphosphate

425. Calmodulin

Questions 426–427

Match each vitamin or cofactor with the correct statement.

(A) Links to the ε-amino group of lysine residues of enzymes
(B) Derives a nitrogen from *p*-aminobenzoic acid
(C) Functions by making and breaking its own disulfide bond
(D) Is a synonym for vitamin B_1
(E) Is a cofactor in deamination of amino acids

426. Folic acid

427. Biotin

Questions 428–430

Each disorder below is associated with a deficiency of a particular vitamin. Match the disorder to the vitamin.

(A) Ascorbic acid (vitamin C)
(B) Thiamine (vitamin B_1)
(C) Riboflavin (vitamin B_2)
(D) Niacin (vitamin B_3)
(E) Biotin
(F) Pantothenic acid
(G) Folic acid
(H) Cobalamin (vitamin B_{12})

428. Wernicke-Korsakoff syndrome associated with alcoholism

429. Spina bifida

430. Pernicious anemia due to lack of intestinal intrinsic factor

Vitamins and Hormones

Answers

361. The answer is E. *(Stryer, 4/e, pp 595, 605–606.)* In humans, both glucagon and epinephrine have been demonstrated to be capable of stimulating adenosine 3′,5′-cyclic monophosphate (cyclic AMP) production in liver, whereas only epinephrine or norepinephrine has been unequivocally shown to be an agonist of cyclic AMP accumulation in adipocytes. In rat adipose tissue, glucagon and adrenocorticotropic hormone (ACTH), in addition to the catecholamines, also stimulate adenylate cyclase. The mode of hormone-receptor stimulation of adenylate cyclase by epinephrine in both adipocytes and hepatocytes is thought to be basically the same. The primary difference between the two tissues lies in the ultimate physiologic response to the accumulation of second messenger and the mode of amplification. In the liver, cyclic AMP activates a protein kinase, which activates a phosphorylase kinase, which finally activates glycogen phosphorylase to turn on glycogenolysis. The cascade pathway is much simpler in adipose tissue, where cyclic AMP activates a protein kinase, which activates triglyceride lipase to turn on lipolysis of stored triacylglycerides.

362. The answer is D. *(Stryer, 4/e, pp 706–707.)* A tumor of the adrenal cortex would be expected to affect the production of adrenal steroid hormone. Cortisol and aldosterone are synthesized in the cortex. In Cushing's syndrome, hypersecretion of cortisol occurs. It is a glucocorticoid that has the effect of encouraging metabolism of proteins, lipids, and carbohydrates. In some cases of Cushing's disease, excessive production of cortisol is due to high levels of ACTH produced as a result of pituitary tumors. Diseases affecting the adrenal medulla might be expected to disrupt or potentiate production of epinephrine in some way. Epinephrine (adrenalin) is synthesized in the medulla.

363. The answer is B. *(Stryer, 4/e, pp 505, 597, 779–780.)* In insulin-dependent diabetics, appropriate insulin injections prevent high blood glucose levels (hyperglycemia) and spilling of glucose into the urine as kidney threshold levels are exceeded (glucosuria). However, insulin is normally injected when blood glucose levels can be maintained by dietary or stored insulin. If insulin injections are given when blood glucose levels are low, hypoglycemia may result. When blood glucose levels fall below 80 mg/dL, insulin shock can occur. Hypoglycemia of 20 mg/dL or lower results in deprivation of glucose to the brain and consequent convulsion. Since the woman in the question did not take in any glucose following an overnight fast, it is possible that her insulin shot will result in hypoglycemia. Ingestion of sugar would prevent any severe consequences. Glucose injections would be necessary if insulin shock occurred. Glycogenolysis and lipolysis are metabolic pathways decreased in the presence of elevated insulin levels. Toxic shock syndrome is a disease not associated with insulin levels.

364. The answer is D. *(Stryer, 4/e, pp 774–780.)* There are two phases of insulin release during glucose stimulation: an acute release that immediately follows glucose stimulation and depends on readily accessible presynthesized insulin for release, and a secondary phase, with a more protracted release that involves mobilization of stored granules of insulin. The secondary phase provides a mechanism to prevent hyperglycemia during periods of prolonged glucose intake.

365. The answer is C. *(Stryer, 4/e, pp 340–341.)* The methylxanthine caffeine (1,3,7-trimethylxanthine), which is found in coffee, is a powerful inhibitor of cyclic nucleotide phosphodiesterase. Thus the caffeine ingested from a cup of coffee would be expected to act synergistically with any hormones that use cyclic AMP as a second messenger and enhance their activity by preventing normal hydrolysis of cyclic AMP.

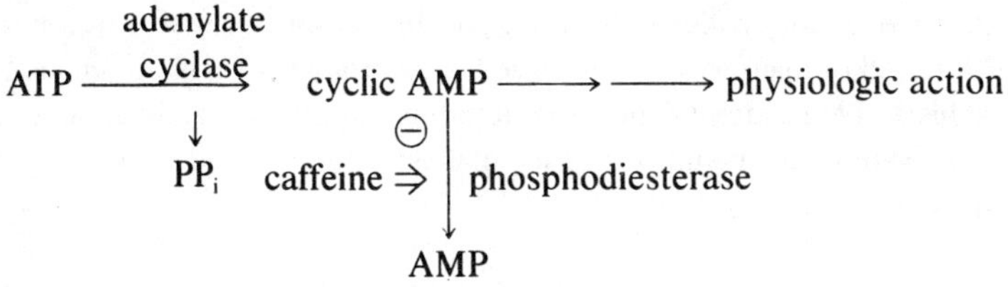

$$\text{ATP} \xrightarrow[\downarrow\ \text{PP}_i]{\text{adenylate cyclase}} \text{cyclic AMP} \longrightarrow \longrightarrow \text{physiologic action}$$

$$\text{cyclic AMP} \xrightarrow[\text{phosphodiesterase}]{\ominus\ \text{caffeine} \Rightarrow} \text{AMP}$$

366. The answer is C. *(Stryer, 4/e, pp 69, 341.)* Vasopressin, which is also called the antidiuretic hormone, increases the permeability of the collecting ducts and distal convoluted tubules of the kidney and thus allows passage of water. Like the mineralocorticoid aldosterone, vasopressin results in an expansion of blood volume. However, the mode of action of aldosterone is different. It causes sodium reabsorption, not water reabsorption. Sodium reabsorption indirectly leads to increased plasma osmolality and, hence, water retention in the blood. Cortisol is a glucocorticoid that potentiates catabolic metabolism chronically. Epinephrine stimulates catabolic metabolism acutely. Insulin acutely favors anabolic metabolism, in large part by allowing glucose and amino acid transport into cells.

367. The answer is B. *(Stryer, 4/e, pp 567–569.)* The antioxidant activity of glutathione is dependent upon maintenance of its reduced state. The enzyme glutathione reductase transfers electrons from NADPH via FAD to oxidized glutathione. Oxidized glutathione is composed of two glutathione molecules held together by a disulfide bridge. Reduced glutathione is a tripeptide with a free sulfhydryl group. It is the presence of the free sulfhydryl group that is of importance to the antioxidant activity of glutathione. In red blood cells, the function of cysteine residues of hemoglobin and other proteins is maintained by the reducing power of glutathione.

368. The answer is D. *(Stryer, 4/e, pp 773–780.)* The glucose tolerance test illustrated in the question shows the typical lag curve of delayed insulin response followed by a hypoglycemic period occurring in 2 to 3 h, or later. This result may be associated with adiposity and early diabetes. In a normal glucose tolerance curve, the peak of blood glucose values occurs between 30 and 45 min following glucose ingestion, and only a slight hypoglycemia or overshoot is recorded.

369. The answer is B. *(Stryer, 4/e, pp 162–163.)* Ferrous iron (Fe^{2+}) is the form absorbed in the intestine by ferritin, transported in plasma by transferrin, and stored in the liver in combination with ferritin or as hemosiderin. There is no known excretory pathway for iron. For this reason, excessive iron uptake over a period of many years may cause hemochromatosis. This is a condition of extensive hemosiderin deposition in the liver, pancreas, and adrenals. The resulting symptoms include cirrhosis, diabetes mellitus, and changes in skin pigmentation.

370. The answer is C. *(Stryer, 4/e, pp 775–777.)* Certain amino acids and lipids are dietary necessities because humans cannot synthesize them. The energy usually obtained from carbohydrates can be obtained from lipids and the conversion of some amino acids to intermediates of the citric acid cycle.

371. The answer is C. *(Stryer, 4/e, pp 642–644.)* Pernicious anemia results from an inability to absorb vitamin B_{12} from the gastrointestinal tract. This may be due to a deficiency of intrinsic factor, surgical gastrectomy, or small bowel disease. The earliest clinical signs of pernicious anemia do not appear until 3 to 5 years following the onset of vitamin B_{12} deficiency.

372. The answer is B. *(Stryer, 4/e, p 707.)* *Osteomalacia* is the name given to the disease of bone seen in adults with vitamin D deficiency. It is analogous to rickets, which is seen in children with the same deficiency. Both disorders are manifestations of defective bone formation.

373. The answer is E. *(Stryer, 4/e, pp 454–455.)* Ascorbic acid (vitamin C) is found in fresh fruits and vegetables. Deficiency of ascorbic acid produces scurvy, the "sailor's disease." Ascorbic acid is necessary for the hydroxylation of proline to hydroxyproline in collagen, a process required in the formation and maintenance of connective tissue. The failure of collagen formation by mesenchymal cells causes the skeletal, dental, and connective tissue deterioration seen in scurvy. Thiamine, niacin, cobalamin, and pantothenic acid can all be obtained from fish or meat products.

374. The answer is C. *(Stryer, 4/e, pp 961, 1027–1029.)* The retinal pigment rhodopsin is composed of the 11-*cis*-retinal form of vitamin A coupled to opsin. Light isomerizes 11-*cis*-retinal to all-*trans*-retinal, which is hydrolyzed to free all-*trans*-retinal and opsin. In order for regeneration of rhodopsin to occur, 11-*cis*-retinal must be regenerated. This dark reaction involves the isomerization of all-*trans*-retinal to 11-*cis*-retinal, which combines with opsin to reform rhodopsin. A deficiency of vitamin A, which is often derived from the β-carotene of plants, results in night blindness.

375. The answer is B. *(Stryer, 4/e, p 754.)* Pellagra is a disease resulting from deficiency of the vitamin niacin (nicotinic acid). Niacin is a component of NAD and NADP. The clinical syndrome characteristic of pellagra, which can include dermatitis, stomatitis, diarrhea, and dementia, actually may result from deficiencies of other nutrients in addition to niacin.

376. The answer is B. *(Stryer, 4/e, pp 570–573.)* Pyruvate carboxylase catalyzes the conversion of pyruvate to oxaloacetate in gluconeogenesis.

$$\text{Pyruvate} + HCO_3^- + \text{ATP} \rightarrow \text{oxaloacetate} + \text{ADP} + P_i$$

In order for pyruvate carboxylase to be ready to function, it requires biotin, Mg^{2+}, and Mn^{2+}. It is allosterically activated by acetyl CoA. The biotin will not be carboxylated until acetyl CoA binds the enzyme. By this means, high levels of acetyl CoA signal the need for more oxaloacetate. When ATP levels are high, the oxaloacetate will be consumed in gluconeogenesis. When ATP levels are low, the oxaloacetate will enter the citric acid cycle. Gluconeogenesis only occurs in the liver and kidneys.

377. The answer is C. *(Stryer, 4/e, p 518.)* Both beriberi and Wernicke's disease are thought to result from deficiency of thiamine (vitamin B_1). Birds manifest thiamine deficiency with opisthotonos. In the United States, thiamine deficiency almost invariably is associated with chronic alcoholism and involves intraventricular brain hemorrhage with respiratory and cardiovascular dysfunction.

378. The answer is B. *(Stryer, 4/e, pp 452–453, 754–755.)* Pantothenic acid, also called *co-acetylase,* is a component of coenzyme A (CoA). Acetyl CoA is the activated form of acetate employed in acetylation reactions, including the citric acid cycle and lipid and cholesterol metabolism. A deficiency of pantothenic acid would limit CoA and have deadly consequences in mammals. However, since it is common in foodstuffs, there is little evidence of pantothenic acid deficiency in humans.

379. The answer is D. *(Stryer, 4/e, pp 255–256.)* In order to be converted to thrombin during clot formation, prothrombin must bind Ca^{2+}, which allows it to anchor to platelet membranes produced by injury. Prothrombin's affinity for Ca^{2+} is dependent on the presence of 10 γ-carboxyglutamate residues found in the first 35 amino acid residues of its amino terminal region. The vitamin K–dependent γ-carboxylation of prothrombin is a posttranslational modification that occurs as nascent prothrombin is synthesized on liver rough endoplasmic reticulum and passes into the lumen of the reticulum. The anticoagulants warfarin and dicumarol are structural analogues that block the γ-carboxylation of prothrombin by substituting for vitamin K. Hence, the prothrombin produced has a weak affinity for Ca^{2+} and cannot properly bind to platelet membranes in order to be converted to thrombin. A simplified diagram of the final steps of fibrin clot formation is given below.

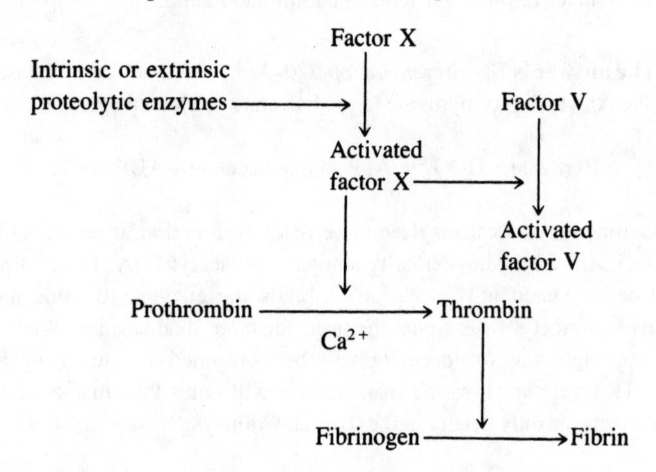

380. The answer is B. *(Stryer, 4/e, pp 572–573, 580, 614–615.)* The vitamin biotin is the cofactor required by carboxylating enzymes such as acetyl CoA, pyruvate, and propionyl CoA carboxylases. The fixation of CO_2 by these biotin-dependent enzymes occurs in two stages. In the first, bicarbonate ion reacts with adenosine triphosphate (ATP) and the biotin carrier protein moiety of the enzyme, and in the second stage, the "active CO_2" reacts with the substrate—e.g., acetyl CoA.

381. The answer is D. *(Stryer, 4/e, pp 706–707.)* Both cortisol and aldosterone have C-21-hydroxyl groups. Both are also derived from progesterone in the adrenal cortex. In contrast, the sex hormones are synthesized in the ovaries and testicular interstitial cells. In the synthesis of sex hormones, progesterone is converted to 17α-hydroxyprogesterone and then androstenedione, which may either become estrone or testosterone. Testosterone gives rise to estradiol in the ovaries. In the corpus luteum, progesterone is produced.

382. The answer is B. *(Stryer, 4/e, pp 754–755.)* Pantothenate is the precursor of CoA, which participates in numerous reactions throughout the metabolic scheme. CoA is a central molecule of metabolism involved in acetylation reactions. Thus a deficiency of pantothenic acid would have severe consequences. There is no documented deficiency state for pantothenate, however, since this vitamin is common in foodstuffs.

383. The answer is C. *(Stryer, 4/e, p 518.)* In the Far East, rice is a staple of the diet. When rice is unsupplemented, beriberi can be manifest, since rice is low in vitamin B_1 (thiamine). Thiamine pyrophosphate is the necessary prosthetic group of enzymes that transfers activated aldehyde units. Such enzymes include transketolase, pyruvate dehydrogenase, and α-ketoglutarate dehydrogenase. Beriberi is a wasting disease whose symptoms include pain in the limbs induced by peripheral neuropathy, weak musculature, and heart enlargement. Yeast products, whole grains, nuts, and pork are rich in thiamine. Choline, ethanolamine, and serine are polar head groups of phospholipids. Glycine is a common amino acid.

384. The answer is C. *(Stryer, 4/e, pp 451– 452, 619–620, 754–755.)* The almost universal carrier of acyl groups is coenzyme A (CoA). However, acyl-carrier protein (ACP) also functions as a carrier of acyl groups. In fatty acid synthesis, ACP carries the acyl intermediates. The reactive prosthetic group of both ACP and CoA is a phosphopantetheine sulfhydryl. In ACP, the phosphopantetheine group is attached to the 77-residue polypeptide chain via a serine hydroxyl. In CoA, the phosphopantetheine is linked to the 5′-phosphate of adenosine that is phosphorylated in its 3′-hydroxyl.

385. The answer is B. *(Stryer, 4/e, pp 449–452, 514–517.)* Lipoamide, like CoA, transfers acetyl groups, but it is a catalytic cofactor in an enzyme complex rather than a stoichiometric cofactor like CoA. A reactive disulfide of lipoamide links the acetyl group to be transferred. Lipoamide becomes acetyllipoamide and then dihydrolipoamide as it first accepts and then transfers an acyl group. This reaction and the regeneration of lipoamide are catalyzed by different parts of the dehydrogenase enzyme complexes of pyruvate dehydrogenase and isocitrate dehydrogenase. ATP transfers phosphoryl groups and thiamine pyrophosphate transfers aldehyde groups. NADH and FADH transfer protons.

386. The answer is D. *(Stryer, 4/e, pp 614–615.)* The key enzymatic step of fatty acid synthesis is the carboxylation of acetyl CoA to form malonyl CoA. The carboxyl of biotin is covalently attached to an ε-amino acid group of a lysine residue of acetyl CoA carboxylase. The reaction occurs in two stages. In the first step, a carboxybiotin is formed.

$$HCO_3^- + \text{biotin-enzyme} + ATP \rightarrow CO_2\text{-biotin-enzyme} + ADP + P_i$$

In the second step, the CO_2 is transferred to acetyl CoA to produce malonyl CoA.

$$CO_2\text{-biotin-enzyme} + \text{acetyl CoA} \rightarrow \text{malonyl CoA} + \text{biotin-enzyme}$$

None of the other cofactors listed are involved in this reaction.

387. The answer is A. *(Stryer, 4/e, pp 346–347, 360, 941–942.)* G_s (stimulatory) and G_i (inhibitory) proteins act as membrane mediators between hormone receptors and the catalytic subunits of adenylate cyclase. Under normal conditions, certain hormones reduce adenylate cyclase activity in target cells, thereby reducing cyclic AMP levels. These hormones, such as α_2-adrenergic amines and opiates, act through their receptors on inhibitory G_i proteins and stimulate hydrolysis of bound GTP, which holds the G protein subunits together. The GDP formed allows the αβγ-subunit complex to split into α- and βγ-subunits. The free βγ-subunits can then bind the α-subunit of G_s proteins, which prevents or reverses activation of adenylate cyclase. In an abnormal parody of this, pertussis toxin blocks the inhibition of adenylate cyclase by ADP-ribosylating G_i protein and preventing GTP hydrolysis. Thus, cyclic

AMP production is not inhibited. Some hormones, such as epinephrine, norepinephrine, and glucagon, occupy receptors and stimulate the activation of G_s by allowing the replacement of GDP with GTP. Abnormally, cholera toxin ADP ribosylates G_s protein and blocks its capacity to hydrolyze GTP to GDP, thereby locking it into an active form, which allows adenylate cyclase to continuously produce cyclic AMP.

388. The answer is B. *(Stryer, 4/e, pp 730, 1001–1002.)* Two of the major hormones are derived from the amino acid tyrosine: the adrenal hormone epinephrine and the thyroid hormone thyroxine (tetraiodothyronine). Epinephrine is the catabolic antagonist of insulin, a polypeptide hormone, and is similar in action to glucagon, a liver-specific, polypeptide hormone. Thyroxine is important in governing the basal metabolic rate.

389. The answer is D. *(Stryer, 4/e, pp 449–453.)* The vitamin whose structure appears in the question is nicotinic acid (niacin), which gives rise to the nicotinamide adenine dinucleotide coenzymes NAD^+ and $NADP^+$. NAD^+ is a cofactor required by all dehydrogenases. NADPH is a cofactor produced by the pentose phosphate shunt. It is utilized in reductive synthesis of compounds such as fatty acids.

390. The answer is B. *(Stryer, 4/e, pp 346–347, 360, 941–942.)* Cholera toxin is an 87-kD protein produced by *Vibrio cholerae,* a gram-negative bacterium. The toxin enters intestinal mucosal cells by binding to G_{m1} ganglioside. It interacts with G_s protein that stimulates adenylate cyclase. By ADP-ribosylation of G_s, the toxin blocks its capacity to hydrolyze bound GTP to GDP. Thus, the G protein is locked in an active form and adenylate cyclase stays irreversibly activated. Under normal conditions, inactivated G protein contains GDP, which is produced by a phosphatase catalyzing the hydrolysis of GTP to GDP. When GDP is so bound to the G protein, the adenylate cyclase is inactive. Upon hormone binding to the receptor, GTP is exchanged for GDP and the G protein is in an active state, allowing adenylate cyclase to produce cyclic AMP. Because cholera toxin prevents the hydrolysis of GTP to GDP, the adenylate cyclase remains in an irreversibly active state, continuously producing cyclic AMP in the intestinal mucosal cells. This leads to a massive loss of body fluid into the intestine within a few hours.

391. The answer is A. *(Stryer, 4/e, pp 745–752.)* The structure shown in the question is the vitamin folic acid. Tetrahydrofolic acid, the active cofactor derived from folic acid, is required in two steps of purine synthesis and thus required in the de novo synthesis of ATP and GTP. CTP and TTP are pyrimidine base derivatives, and although de novo synthesis of the pyrimidine ring does not require tetrahydrofolate, the formation of thymine from uracil does. NADH and NADPH require niacin for their synthesis.

392. The answer is C. *(Stryer, 4/e, pp 449–453, 754–755.)* The structure shown in the question is that of the vitamin riboflavin. It is a precursor of two cofactors involved in electron transport systems, riboflavin 5′-phosphate, also known as *flavin mononucleotide (FMN),* and flavin adenine dinucleotide (FAD). Strictly speaking, these compounds are not nucleotides, as they contain the sugar alcohol ribitol, not ribose. The cofactors are strongly bound to their apoenzymes and function as dehydrogenation catalysts. Pyruvate dehydrogenase is a multienzyme complex and contains the enzyme dihydrolipoyl dehydrogenase, which has as its prosthetic group two molecules of FAD per molecule of enzyme. In the overall reaction, the reduced FAD is reoxidized by NAD^+. Succinate dehydrogenase also contains tightly bound FAD, one molecule per molecule of enzyme. Glutamate, lactate, malate, and glyceraldehyde-3-phosphate dehydrogenases all use nicotinamide dinucleotide cofactors and do not contain FAD as a prosthetic group.

393. The answer is B. *(Stryer, 4/e, pp 621–622.)* Acetyl CoA carboxylase is the major regulatory enzyme of fatty acid synthesis. The regulation of this enzyme is quite similar to that of the regulation of glycogen metabolism in liver; that is, insulin promotes the dephosphorylation of acetyl CoA carboxylase, which leads to its activation. In contrast, glucagon leads to the phosphorylation and deactivation of acetyl CoA carboxylase through a cyclic AMP amplification mechanism. Acetyl CoA carboxylase can be compared with glycogen synthase with respect to effects of glucagon and insulin on its phosphorylation-dephosphorylation regulation.

394. The answer is C. *(Stryer, 4/e, pp 343–346.)* Upon appropriate hormonal stimulation, activation of phospholipase C (phosphoinositidase) leads to the hydrolysis of phosphatidylinositol 4,5-bisphosphate (PIP_2) in the plasma membrane. Two messengers, diacylglycerol and inositol 1,4,5-triphosphate (IP_3), are released. IP_3 causes calcium channels in the endoplasmic reticulum to open and release calcium from intracellular stores. Calcium ion is also an intracellular messenger that affects a variety of cellular processes. These include smooth muscle contraction, aggregation of blood platelets, insulin secretion by pancreatic islet cells, and histimine secretion by mast cells, to name a few. Calcium seems to cause its effects through modulation of calcium-binding proteins such as calmodulin.

395. The answer is E. *(Stryer, 4/e, pp 452–453, 630–634.)* The coenzyme pyridoxal phosphate is a versatile compound that aids in the catalysis of many reactions involving amino acids. This includes amino transferase reactions such as transamination (e.g., glutamic-oxaloacetic transaminase), deamination (e.g., serine dehydratase), decarboxylation (glutamate decarboxylase), and transulfuration (e.g., cystathionine synthetase and cystathionase). Using the same type of mechanism, pyridoxal phosphate is important for the operation of glycogen phosphorylase. All of these reactions using pyridoxal phosphate catalysis have a number of features in common. First, a Schiff-base intermediate is formed with a specific lysine group at the active site of the appropriate enzymes. Then, in the reactions involving amino acids, the amino acid substrate is exchanged to form the Schiff-base with the amino acid substrate. Finally, the proteinated form of pyridoxal phosphate acts as an electron sink to stabilize the catalytic intermediates that are negatively charged. In other words, the ring nitrogen of pyridoxal phosphate attracts electrons from the amino acid substrate, which allows the product Schiff-base to be hydrolyzed. Pyridoxine (vitamin B_6) is found in many different foods. Therefore, deficiency usually only results from the administration of a number of commonly used drugs that act as pyridoxine antagonists (e.g., isoniazid and penicillamine).

396. The answer is A. *(Stryer, 4/e, pp 559–560, 754–755.)* The major contributor of electrons in reductive biosynthetic reactions is nicotinamide adenine dinucleotide phosphate (NADPH + H^+), which is derived by reduction of NAD^+. NAD^+ is formed from the vitamin niacin (also called *nicotinate*). Niacin can be formed from tryptophan in humans. In the synthesis of NAD^+, niacin reacts with 5-phosphoribosyl-1-pyrophosphate to form nicotinate ribonucleotide. Then, AMP is transferred from ATP to nicotinate ribonucleotide. Finally, the amide group of glutamate is transferred to the niacin carboxyl group to form the final product, NAD^+. $NADP^+$ is derived from NAD^+ by phosphorylation of the 2′ hydroxyl group of the adenine ribose moiety. The reduction of $NADP^+$ to NADPH + H^+ occurs primarily through the hexose monophosphate shunt.

397. The answer is C. *(Stryer, 4/e, pp 449–455.)* A coenzyme is a nonprotein organic molecule that binds to an enzyme to aid in its catalytic function. Usually it is involved in the transfer of a specific functional group. A coenzyme usually binds loosely and can be separated from the enzyme. When a coenzyme binds tightly to an enzyme, it is spoken of as a *prosthetic group* of the enzyme. While many coenzymes are derived by modification of vitamins such as CoA from pantothenic acid or FAD from riboflavin, some coenzymes do not contain vitamins. For example, ubiquinone (oxidized) or ubiquinol (reduced) is, in fact, coenzyme Q, which is involved in transferring hydrogen ion atoms and electrons in the oxidative phosphorylation system. Since humans can synthesize ubiquinones, they are not considered vitamins. Likewise, lipoic acid is a short-chain fatty acid with two sulfhydryl groups. Lipoic acid is involved in pyruvate dehydrogenase and α-ketoglutarate dehydrogenase reactions. Each of these reactions also uses thiamine pyrophosphate, CoA, FAD^+, and NAD^+. Tetrahydrofolic acid is derived from folic acid, a B vitamin that is obtained in the diet. Folic acid is modified by cleavage of a chain of glutamate residues that yields tetrahydrofolic acid, which is involved in carrying out 1-carbon transfer reactions. Since coenzymes are not protein molecules, pyruvate dehydrogenase cannot be considered a coenzyme.

398. The answer is D. *(Stryer, 4/e, pp 452–455.)* Cofactors are distinguished from coenzymes because cofactors are usually metallic ions instead of organic molecules. Examples of cofactors include copper in cytochrome oxidase, iron in all the cytochromes, magnesium for all enzymes utilizing ATP, and zinc in lactate dehydrogenase. Methylcobalamin is obtained from the modification of cyanocobalamin (vitamin B_{12}). Thus, methylcobalamin is not a cofactor.

399. The answer is E. *(Stryer, 4/e, pp 255–256.)* The major role of vitamin K is in the synthesis of prothrombin and other clotting factors (e.g., VII, IX, and X). Vitamin K acts on the inactive precursor molecules of these proteins, allowing carboxylation of glutamic acid residues to γ-carboxyglutamate. A true vitamin K deficiency is unusual because vitamin K is found in a variety of foods and can be produced by intestinal bacteria. Liver, egg yolk, spinach, cauliflower, and cabbage are some of the sources of vitamin K. Vitamin K is not involved in the prevention of thrombosis (blood clots that can lead to stroke or heart failure), but in fact is required for the formation of blood clots. In fact, a method for preventing thrombosis involves the use of drugs that interfere with vitamin K, such as dicumarol and warfarin, a synthetic analogue of vitamin K. Both of these compounds interfere with the formation of γ-carboxyglutamate.

400. The answer is C. *(Stryer, 4/e, pp 333–335, 1002.)* Yellow and dark green vegetables as well as fruits are good sources of carotenes, which serve as precursors of vitamin A. However, egg yolk, butter, cream, and liver and kidneys are good sources of preformed vitamin A. The visual pigment rhodopsin is formed from the protein opsin and 11-*cis*-retinal. During the photobleaching of rhodopsin, all-*trans*-retinal plus opsin is formed from dissociated rhodopsin, causing an impulse that is transmitted by the optic nerve to the brain. 11-*Cis*-retinal is isomerized from *trans* retinal, which spontaneously combines with opsin to reform rhodopsin, making it ready for another photochemical cycle. Vitamin A is essential for the normal differentiation of epithelial tissue as well as normal reproduction. All-*trans* retinoic acid (tretinoin) has been found to be effective for topical treatment of psoriasis. Another form of vitamin A is 13-*cis*-retinoic acid, which has been found to be effective in the treatment of severe cases of acne. Vitamin A is not synthesized in the skin. Vitamin D can be synthesized in the skin under the influence of sunlight from 7-dehydrocholesterol, an intermediate in cholesterol synthesis.

401. The answer is D. *(Stryer, 4/e, pp 351–352, 773–775.)* Epidermal growth factor receptor is similar to the insulin receptor in that it has a tyrosine kinase activity that is activated by the binding of growth factor to the extracellular portion of the protein. Neither insulin nor epidermal growth factor is known to lead to hyperglycemia. In fact, insulin leads to hypoglycemia by increasing movement of glucose into cells. In contrast, epinephrine and glucagon lead to acute hyperglycemic effects by activating liver phosphorylase and the release of glucose through a cyclic AMP cascade effect. Glucocorticoid, a steroid hormone, and thyroxine have chronic effects and act by binding to intracellular binding proteins (receptors) that eventually act as enhancers of transcription. The mechanism by which thyroid hormones raise serum glucose levels is unknown. Glucocorticoids stimulate the liver to produce more gluconeogenic enzymes and promote protein breakdown to form amino acids. In this manner, glucocorticoids promote a hyperglycemic effect.

402. The answer is C. *(Stryer, 4/e, pp 255–256.)* Vitamin K is essential for the posttranscriptional modification of prothrombin by γ-carboxylation of glutamate residues. A functional deficiency exists in patients treated with analogues of vitamin K such as the coumarin derivatives. The analogues act as anticoagulants by competing with vitamin K and preventing the production of functional prothrombin. By administration of vitamin K, hemorrhage can be prevented in such patients. Vitamin K is normally obtained from green, leafy vegetables in the diet. In addition, intestinal bacteria synthesize the vitamin. Hence, lack of green, leafy vegetables or prolonged antibiotic therapy that sterilizes the gut can lead to a deficiency. A deficiency of vitamin K can cause hemorrhage disease in newborn infants since their intestines do not have the bacteria that produce it and since vitamin K does not cross the placenta. The absence of red meat from the diet should not affect the status of vitamin K in the body.

403. The answer is B. *(Stryer, 4/e, pp 25, 773–774, 779–780.)* The anabolic hormone called insulin is synthesized on the endoplasmic reticulum of pancreatic β-cells as a nascent polypeptide chain called *preproinsulin.* Immediately following synthesis, the amino-terminal signal sequence of 16 residues is cleaved off to form proinsulin. Proinsulin is composed of one continuous polypeptide that contains in sequence an A chain of 21 residues, a connecting peptide (C peptide) of about 30 residues, and a B chain of 30 residues. The molecule is folded so that two disulfide bridges span the A and B chains. The proinsulin molecule is transported from the lumen of the endoplasmic reticulum to the Golgi apparatus, where it is packaged into storage granules. In the Golgi apparatus and in the storage granules, proteolysis of the C peptide occurs. Exocytosis of the granules releases insulin as well as C peptides into the bloodstream. Neither proinsulin nor the C peptide is biologically active. The action of insulin is antagonistic to glucagon, which is a catabolic hormone secreted by the α-cells of the pancreas.

404. The answer is D. *(Stryer, 4/e, pp 340–343.)* Many polypeptide hormones exert their effects upon target tissues through the second messenger, cyclic AMP. These include glucagon, which acts specifically on liver to catabolically raise blood glucose levels, as well as epinephrine, which induces glycogenolysis in muscle and lipolysis in adipose tissues. Follicle-stimulating hormone and luteinizing hormone are secreted by the anterior pituitary. Follicle-stimulating hormone maintains the structure of the seminiferous tubules, while luteinizing hormone (also known as interstitial cell hormone) activates interstitial cells of the testis to produce testosterone. In females, follicle-stimulating hormone supports growth of the ovarian follicle up to the point of ovulation. The developing follicle secretes estrogens, which stimulate development of the endometrium. Luteinizing hormone, together with follicle-stimulating hormone, is required for ovulation and early development of the corpus luteum. The corpus luteum secretes progesterone, which maintains the endometrium. Estrogen and progesterone are steroid hormones that, unlike polypeptide hormones, penetrate into cell nuclei to activate specific genes.

405. The answer is C. *(Stryer, 4/e, pp 773–781.)* Insulin allows the disposition and utilization of glucose—particularly exogenous glucose. High blood glucose signals pancreatic β-cells to secrete insulin. Under these conditions, insulin stimulates the entry of glucose and amino acids into a variety of tissues, including muscle and fat cells. The presence of glucose allows the anabolic processes of glucose oxidation, lipogenesis, and the synthesis of macromolecular precursors, such as nucleotides, to be carried out. In contrast, gluconeogenesis is a catabolic process for the synthesis of glucose, mainly from the amino acids of degraded proteins. Gluconeogenesis is the adaptive response of the organism to low blood levels of glucose and is, therefore, diminished by insulin.

406. The answer is A. *(Stryer, 4/e, pp 773–774.)* The actions of epinephrine (adrenalin) and norepinephrine are catabolic; that is, these catecholamines are antagonistic to the anabolic functions of insulin and, like glucagon, are secreted in response to low blood glucose or during "fight-or-flight" stress. Unlike glucagon, which only acts on liver, the catecholamines affect most tissues, including liver and muscle. The catabolic processes increased by secretion of epinephrine and norepinephrine include glycogenolysis, gluconeogenesis, lipolysis, and ketogenesis. Thus, products that increase blood sugar or spare it, such as ketone bodies and fatty acids, are increased. Glycolysis is an anabolic process that is decreased in the presence of elevated catecholamines.

407. The answer is D. *(Stryer, 4/e, pp 622–624.)* Although prostaglandins were originally isolated from prostate glands, seminal vesicles, and semen, their synthesis in other organs has been amply documented; indeed, few organs have failed to demonstrate prostaglandin release. Although prostaglandins are not completely understood, it seems likely that they function as local or intracellular signals. Platelet aggregation, smooth muscle contraction, vasodilation, and uterine contraction are but a few of the functions these potent, cyclic acidic lipids may regulate.

408. The answer is E. *(Stryer, 4/e, pp 544–549, 607–608.)* Carnitine increases the transport of fatty acids into mitochondria and thus stimulates their oxidation. A patient with systemic deficiency of carnitine may have severe hypoglycemia: If the peripheral tissues cannot utilize fatty acids for production of energy, all tissues become glucose-dependent. The other compounds listed in the question are members of the transport chain; they convey electrons to oxygen in the mitochondrial respiratory chain.

409. The answer is A. *(Stryer, 4/e, pp 69, 691–692, 696, 702–709, 993.)* Adrenocorticotropin (ACTH) is a peptide hormone of the adenohypophysis that influences the secretion of corticosteroid hormones. The other compounds listed in the question contain the basic steroid-ring system and are ultimate derivatives of cholesterol. The compound whose structure appears in the question is cholesterol, one of a large group of steroids. Cholesterol, which can be derived from the diet as well as synthesized de novo, is the precursor of all steroids involved in mammalian metabolism. These include the bile acids, the steroid hormones, and vitamin D. Cholesterol cannot be metabolized to carbon dioxide and water in humans. It must be excreted as a component of the bile.

410. The answer is A. *(Stryer, 4/e, pp 514–517.)* Pyruvate dehydrogenase is an organized enzyme assembly containing pyruvate dehydrogenase, dihydrolipoyl transacetylase, dihydrolipoyl dehydrogenase, and two enzymes involved in regulation of the overall enzymatic activity of the complex. Pyruvate dehydrogenase requires thiamine pyrophosphate as its cofactor, dihydrolipoyl transacetylase requires lipoic acid and CoA, and dihydrolipoyl dehydrogenase has an FAD prosthetic group that is reoxidized by NAD^+. The only cofactor listed in the question that is not required in the pyruvate dehydrogenase reaction is pyridoxal phosphate, a cofactor widely used by enzymes involved in amino acid metabolism, such as the transaminases.

411. The answer is A. *(Stryer, 4/e, pp 341–343.)* Hormones are molecules synthesized by glandular tissues and secreted directly into the bloodstream. They alter the activities of specific, responsive target issues. As such, hormones act by regulating preexisting processes. They may influence the synthesis of proteins or enzymes (e.g., steroid hormones), affect the rate of enzymatic catalysis (e.g., cyclic AMP–mediated hormones), or alter the permeability of plasma membranes (e.g., insulin). To date, no hormone that acts as a vitamin has been observed.

412. The answer is D. *(Stryer, 4/e, pp 452–453, 754–755.)* Riboflavin (vitamin B_2) functions in the coenzyme forms of flavin mononucleotide (FMN) or flavin adenine dinucleotide (FAD). When concentrated, both have a yellow color due to the riboflavin. Both function as prosthetic groups of oxidation-reduction enzymes or flavoproteins. They transfer electrons and protons. Thiamine occurs functionally as thiamine pyrophosphate, which is also known as *cocarboxylase.* Nicotinamide adenine dinucleotide (NAD) is the functional coenzyme derivative of niacin. It is the major electron acceptor in the oxidation of molecules.

413. The answer is B. *(Stryer, 4/e, pp 449–453, 514–517.)* All the vitamins listed except lipoamide contain at least one phosphate in their cofactor form. Thiamine (vitamin B_1) is converted to thiamine pyrophosphate simply by the addition of pyrophosphate. It is involved in aldehyde group transfer. Niacin (nicotinic acid) is esterified to adenine dinucleotide and its two phosphates to form nicotinamide adenine dinucleotide. Pyridoxine (vitamin B_6) is converted to either pyridoxal phosphate or pyridoxamine phosphate before complexing with enzymes. Riboflavin becomes flavin mononucleotide by obtaining one phosphate (riboflavin 5′-phosphate). If it complexes with adenine dinucleotide via a pyrophosphate ester linkage, it becomes flavin adenine dinucleotide.

414. The answer is C. *(Stryer, 4/e, pp 513–514.)* The net result of the citric acid cycle's oxidation of acetyl CoA is shown below.

$$\text{Acetyl CoA} + \text{FAD} + 3\ \text{NAD}^+ + \text{GDP} + 2\ \text{H}_2\text{O} + \text{P}_i \rightarrow 2\ \text{CO}_2 + \text{CoA} + \text{FADH}_2 + 3\ \text{NADH} + \text{GTP} + 2\ \text{H}^+$$

Two carbon atoms enter the cycle as acetyl CoA, with an immediate loss of CoA as citrate is formed from oxaloacetate. Two carbon atoms leave the cycle at the level of isocitrate dehydrogenase and α-ketoglutarate dehydrogenase. The two carbons that leave as CO_2 are actually not the original acetyl CoA carbons. Two NAD^+ molecules are reduced by isocitrate dehydrogenase and then α-ketoglutarate dehydrogenase. One FAD is reduced during the oxidation of succinate, and one NAD^+ is reduced when malate is oxidized. GTP is formed from GDP by utilization of the high-energy thioester linkage of succinyl CoA.

415. The answer is B. *(Stryer, 4/e, pp 452–454.)* Vitamins A, D, E, and K are all fat-soluble. Thus, by definition, they would be soluble in organic solvents used to extract lipids. The physical characteristics of fat-soluble vitamins derive from the hydrophobic nature of the aliphatic chains composing them. In contrast, all the other vitamins known are classified as water-soluble.

416. The answer is E. *(Stryer, 4/e, pp 451–452.)* Coenzyme A (CoA) is one of the major molecules in metabolism. The *A* stands for *acetylation.* Acetyl groups are linked to the reactive terminal sulfhydryl group to produce acetyl CoA, which has a high acetyl-transfer potential. CoA carries and transfers acetyl groups in much the same way as activated phosphoryl groups are transferred by ATP. However, CoA is not involved in pathways in which acetyl transfer does not occur, such as glycolysis.

417. The answer is C. *(Stryer, 4/e, pp 773–775.)* Both glucagon and epinephrine are antagonists of insulin. They raise blood glucose levels by stimulating glycogenolysis via cyclic AMP mediation. Glucagon acts specifically on the liver and kidneys, whereas epinephrine can stimulate these tissues as well as skeletal muscle and adipocytes. The 11-hydroxy, C-21 adrenocortical steroids known as *glucocorticoids* promote increased blood levels by being permissive to the actions of glucagon and epinephrine; that is, they promote protein and amino acid degradation to oxaloacetate and pyruvate, which are substrates for gluconeogenesis. In addition, glucocorticoids induce liver production of gluconeogenic enzymes. Hence, by acute stimulation as well as by adaptive changes, glycogenolytic hormones and the glucocorticoids promote hyperglycemia. In contrast, insulin promotes glucose uptake by cells to lower blood glucose levels and facilitates amino acid entry into cells, which lowers the amino acid supply available for gluconeogenesis.

418. The answer is E. *(Stryer, 4/e, pp 514–517.)* The vitamin whose structure appears in the question is thiamine (vitamin B_1). Thiamine pyrophosphate is required for the reactions catalyzed by pyruvate dehydrogenase, transketolase, pyruvate decarboxylase, and α-ketoglutarate dehydrogenase. In all these reactions, thiamine is involved with oxidative decarboxylation. Citrate synthetase involves neither oxidative decarboxylation nor thiamine.

419. The answer is B. *(Stryer, 4/e, pp 630–633.)* The structure shown in the question is that of the vitamin pyridoxine (vitamin B_6). The active cofactor derived from pyridoxine is pyridoxal phosphate. All transaminase enzymes and many other enzymes involved in amino acid metabolism require pyridoxal phosphate as a cofactor. The cofactor is also essential for the catalytic activity of glycogen phosphorylase, in which it appears to serve a purely structural function. Dehydrogenases do not require pyridoxal phosphate for activity.

420–421. The answers are 420-E, 421-A. *(Stryer, 4/e, pp 449–451, 514–517, 585–587.)* The activated form of glucose utilized for the synthesis of glycogen and galactose is UDP-glucose, which is formed from the reaction of glucose-1-phosphate and UTP.

The vitamin lipoic acid is covalently bound to the ε-amino group of a lysine residue of the enzyme dihydrolipoyl transacetylase. The amide-linked lipolysine residue is known as *lipoamide* and is an activated carrier of acyl groups derived from the hydroxyethyl derivative of thiamine pyrophosphate. In this manner, lipoamide functions as one of the coenzymes in oxidative decarboxylation reactions.

In reductive synthesis such as fatty acid synthesis, NADPH is the major electron donor. This may be contrasted to NADH, which is utilized for the generation of ATP via electron transport.

422–423. The answers are 422-E, 423-A. *(Stryer, 4/e, pp 247–251.)* Secretin, a circulatory hormone liberated in response to peptides or acid in the duodenum, stimulates the flow of pancreatic juice. Gastrin governs acid production by the stomach, and cholecystokinin causes the gallbladder to contract. Cholecystokinin stimulates this contraction after it is released by the duodenum into the circulation, with subsequent emptying of bile into the intestine. The C-terminal octapeptide of cholecystokinin is more than five times as potent as the parent hormone, and its C-terminal pentapeptide is identical to gastrin. Gastrin, produced in specialized cells of the antral mucosa of the stomach, stimulates parietal cells to produce HCl (approximately 0.16 *M*) and KCl (0.007 *M*); it also stimulates secretion of glucagon and insulin. Production of gastrin is inhibited by secretin.

424–425. The answers are 424-C, 425-E. *(Stryer, 4/e, pp 343–345, 349–350.)* A variety of agonists activate the plasma membrane–bound enzyme phospholipase C, which hydrolyzes the phosphodiester bond of phosphatidyl inositol 4,5-bisphosphate and consequently releases diacylglycerol (DAG) and inositol 1,4,5-triphosphate (IP_3). Phospholipase C is also known as *phosphoinositidase* and as *polyphosphoinositide phosphodiesterase.* Both DAG and IP_3 are second messengers. DAG activates protein kinase C, which is important in controlling cell division and cell proliferation. IP_3 opens calcium channels and allows the rapid release of the calcium stores in endoplasmic reticulum (in smooth muscle, sarcoplasmic reticulum). The elevated levels of calcium ion stimulate smooth muscle contraction, exocytosis, and glycogen breakdown.

In many enzymes affected by calcium ions, calmodulin is the calcium detector. It binds Ca^{2+} at four sites and undergoes a conformational change allowing it to stimulate many enzymes. For example, calmodulin is a subunit of glycogen phosphorylase kinase.

426–427. The answers are 426-B, 427-A. *(Stryer, 4/e, pp 572–573, 630–633, 719–723.)* Folic acid is built from a substituted pteridine, glutamic acid, and *p*-aminobenzoic acid. It is important in the transfer of one-carbon groups.

Biotin is composed of thiophene and imidazole rings and is usually synthesized by intestinal bacteria. When functioning as a cofactor, its side-chain carboxyl is linked to the ε-lysine of enzymes via an amide bond.

Pyridoxine (vitamin B_6) is a cofactor involved in the deamination and decarboxylation of amino acids. In particular, the vitamin is important in the metabolism of sulfur-containing amino acids, glycine, tryptophan, serine, and glutamate. Synthesis of the heme precursor δ-aminolevulinic acid requires pyridoxal phosphate.

428–430. The answers are 428-B, 429-G, 430-H. *(Stryer, 4/e, pp 453, 566–567, 642–643, 729–730, 752–754.)* Severe alcoholics can develop the Wernicke-Korsakoff syndrome, which is characterized by loss of memory, lackadaisical behavior, and a continuous rhythmic movement of the eyeballs. The syndrome is due to a thiamine deficiency caused by an inappropriate diet.

An important breakthrough in the prevention of spina bifida (a neural tube defect at birth) was the discovery that folic acid is absolutely crucial for the development of the neural tube in the first few weeks of fetal life. Since this is a time when many women are not aware that they are pregnant, it is essential that all women of child-bearing age take a folic acid supplement of approximately 0.4 mg per day. Frank folic acid deficiency can also cause megaloblastic anemia because of a decreased synthesis of the purines and pyrimidines needed for cells to make DNA and divide.

A lack of intrinsic factor, which is a glycoprotein secreted by gastric parietal cells, is often the cause of vitamin B_{12} (cobalamin) deficiency. This is not an actual deficiency of vitamin B_{12} in the diet. Vitamin B_{12} binds to the intrinsic factor and only in this state can it bind to specific receptors on the mucosal cells of the ileum and be taken up. In addition to pernicious anemia, lack of vitamin B_{12} can lead to central nervous system defects that are irreversible and lead to neuropsychiatric symptoms. The only known treatment for intrinsic factor deficiency (vitamin B_{12} deficiency) is intramuscular injection of cyanocobalamin throughout the patient's life.

Membranes and Cell Structure

DIRECTIONS: Each question below contains five suggested responses. Select the **one best** response to each question.

431. In order for a reconstituted system of myosin-coated beads to show structured movement in vitro, the appropriate medium must contain ATP and at least

(A) calcium + actin
(B) calcium + actin + tropomyosin
(C) calcium + actin + tropomyosin + troponin
(D) actin + tropomyosin + troponin + microtubules
(E) tropomyosin + troponin

432. A typical mammalian plasma membrane would be most likely to have which of the following approximate weight compositions?

	L	P	C	RNA
(A)	35%	45%	5%	10%
(B)	35%	55%	5%	0%
(C)	35%	40%	20%	0%
(D)	60%	30%	0%	5%
(E)	20%	75%	0%	0%

L = lipid, P = protein, C = carbohydrate.

433. Which of the following compounds is an inhibitor of sodium-dependent glucose transport across the plasma membrane?

(A) Ouabain
(B) Sodium azide
(C) Dicumarol
(D) Phlorhizin
(E) Phloretin

434. Which subcellular fraction of mammalian hepatocytes contains the enzymes necessary for elongation of long-chain fatty acids?

(A) Plasma membrane
(B) Nucleus
(C) Endoplasmic reticulum
(D) Organelle-free cytosol
(E) Ribosomes

Questions 435–436

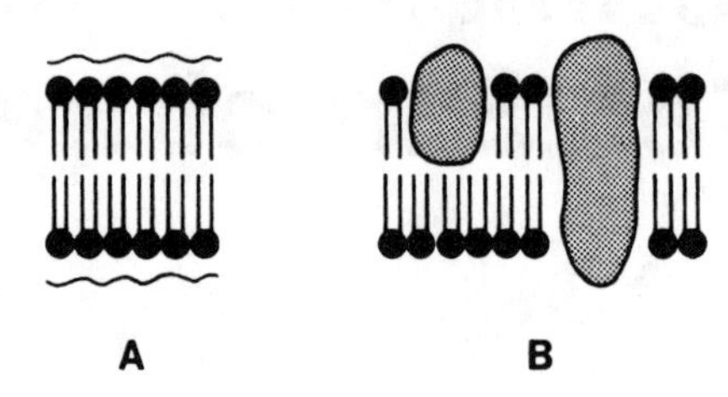

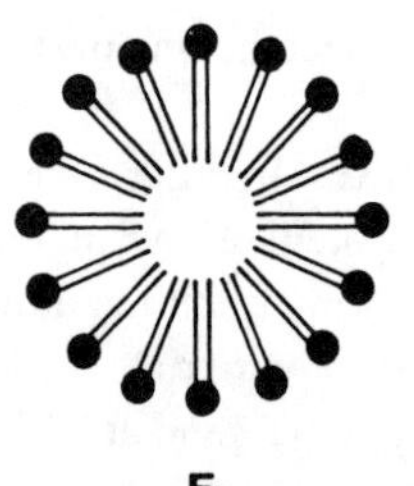

435. Which of the diagrammatic structures shown above represents the model of biologic membranes that most successfully accounts for membrane asymmetry?

(A) A
(B) B
(C) C
(D) D
(E) E

436. Which of the diagrammatic structures shown above most clearly represents a model of the configuration of an artificial bilayer composed solely of phospholipids?

(A) A
(B) B
(C) C
(D) D
(E) E

437. Which one of the following features is predicted by the Nicholson-Singer fluid mosaic model of biologic membranes?

(A) Membrane lipids do not diffuse laterally
(B) Membrane lipid is primarily in a monolayer form
(C) Membrane lipids freely flip-flop
(D) Membrane proteins may diffuse laterally
(E) Membrane proteins are layered at aqueous interfaces

438. New proteins destined for secretion are synthesized in the

(A) Golgi apparatus
(B) smooth endoplasmic reticulum
(C) free polysomes
(D) nucleus
(E) rough endoplasmic reticulum

DIRECTIONS: Each numbered question or incomplete statement below is NEGATIVELY phrased. Select the **one best** lettered response.

439. All the following occur *only* in the structure indicated in the electron micrograph below EXCEPT

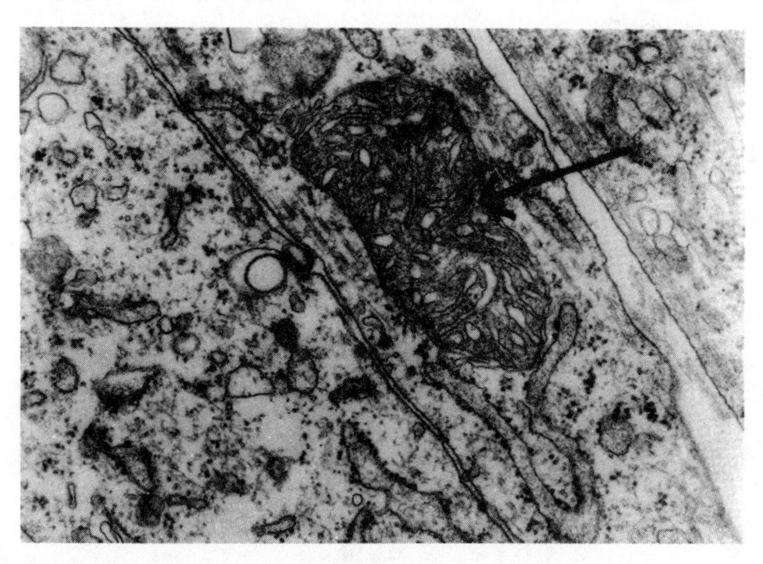

(A) formation of citrate from oxaloacetate and acetyl CoA
(B) reduction of isocitrate to form α-ketoglutarate
(C) oxidative decarboxylation of pyruvate to form acetyl CoA
(D) carboxylation of pyruvate to form oxaloacetate
(E) reversible reduction of malate to form oxaloacetate

---

440. All the following processes occur in the endoplasmic reticulum EXCEPT

(A) urea cycle
(B) fatty acid elongation
(C) protein glycosylation
(D) fatty acid desaturation
(E) secretory protein synthesis

441. All the following components make up the cytoskeleton of most eukaryotic cells EXCEPT

(A) actin filaments
(B) microfilaments
(C) intermediate filaments
(D) chromatin
(E) microtubules

442. The thin section composed of isolated red blood cell ghost membranes illustrated in the electron micrograph below contains all the following components EXCEPT

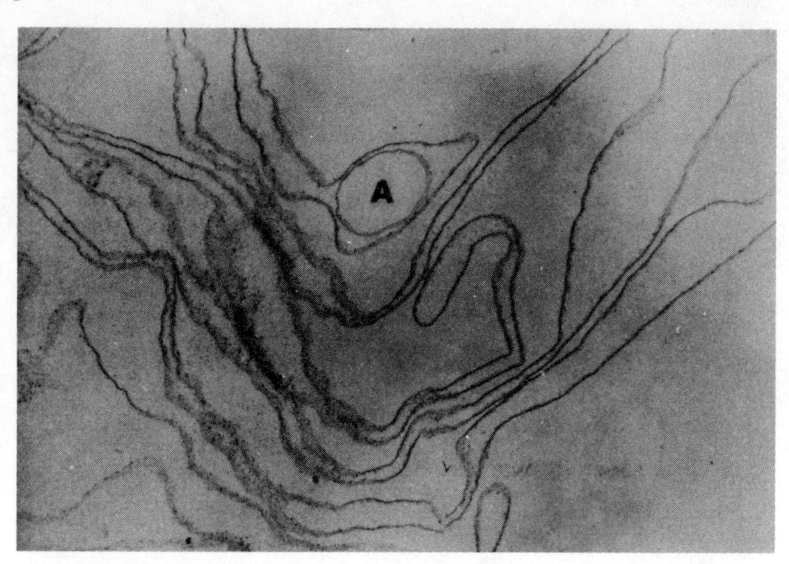

(A) cholesterol
(B) nucleic acid
(C) proteins
(D) carbohydrates
(E) gangliosides

---

443. ATP is the direct source of free energy in all the following active transport systems of mammals EXCEPT

(A) Na^+, K^+
(B) glucose
(C) Ca^{2+}
(D) H^+ of stomach
(E) proton pump of endocytic vesicles

444. All the following enzymes can be isolated as part of a complex of morphologically discrete particles that can be visualized by electron microscopy EXCEPT

(A) glucokinase
(B) glycogen phosphorylase
(C) pyruvate dehydrogenase
(D) mitochondrial ATPase (ATP synthase)
(E) fatty acid synthetase

445. All the following enzymes are found in lysosomes EXCEPT

(A) acid phosphatase
(B) lipoprotein lipase
(C) *N*-acetylhexosaminidase
(D) cathepsin
(E) β-galactosidase

DIRECTIONS: The group of questions below consists of lettered headings followed by a set of numbered items. For each numbered item select the **one** lettered heading with which it is **most** closely associated. Each lettered heading may be used **once, more than once, or not at all.**

Questions 446–448

For each biologic process or molecular structure given, select the corresponding letter designating a structure in the electron micrograph below with which it is most commonly associated. Answer **E** if the process or molecular structure is not associated with any of the lettered sites.

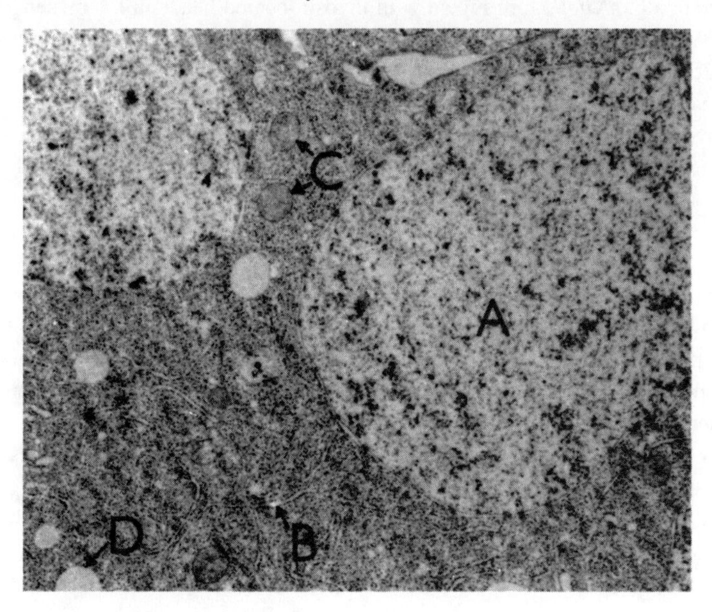

446. Activation of ketone bodies

447. Insulin receptor

448. Desaturation of fatty acids

Membranes and Cell Structure

Answers

431. The answer is C. *(Stryer, 4/e, pp 401–402.)* When the actin cables of a giant alga, *Nitella,* are mixed with myosin-coated beads in the presence of ATP, unidirectional movement of the beads along the cables can be observed under the appropriate conditions. If only calcium is added to the system, no movement is observed. However, if tropomyosin and troponin are also added, movement will occur. By selective addition and deletion it has been determined that calcium is the controlling signal for muscular contraction and that an allosteric flow of reactions proceeds: calcium ions → troponin → tropomyosin → actin → myosin.

432. The answer is B. *(Rawn, pp 222–226. Stryer, 4/e, pp 267–268.)* Although all plasma membranes contain lipid, protein, and carbohydrate, their chemical makeup can vary widely. For example, myelin membranes contain about 75 percent lipid, 20 percent protein, and 5 percent carbohydrate by weight, whereas erythrocyte ghosts are composed of 43 percent lipid, 49 percent protein, and 8 percent carbohydrate. In general, liver cell plasma membranes are considered representative and contain about 39 percent lipid, 54 percent protein, and 7 percent carbohydrate. Any ribonucleic acid (RNA) observed in an isolated preparation of mammalian plasma membranes is considered to be a contaminant and is usually less than 0.1 percent by weight.

433. The answer is A. *(Stryer, 4/e, pp 313–314.)* The digitalis glycoside ouabain inhibits Na^+,K^+-ATPase, which ejects Na^+ from cells. This plasma membrane ATPase is responsible for maintaining the membrane potential and osmotic gradient of cells. Related glycosides include phlorhizin, an inhibitor of sodium-dependent glucose transport, and phloretin, an inhibitor of sodium-independent facilitated diffusion of glucose.

434. The answer is C. *(Stryer, 4/e, pp 622–623.)* Cytoplasmic synthesis of fatty acids results in the production of the 16-carbon fatty acid palmitate. Elongation of this as well as longer fatty acids is carried out by enzymes localized to the endoplasmic reticulum. Similar to the cytoplasmic system's leading to the production of palmitate, malonyl CoA is utilized for 2 carbon additions to the elongating chain. Double bonds also are inserted by enzymes present in the endoplasmic reticulum.

435. The answer is B. *(Stryer, 4/e, pp 268–276.)* The fluid mosaic model of membrane structure shown in the question describes plasma membranes as a mosaic of globular proteins in a phospholipid bilayer. The lipids as well as the proteins are in a fluid and dynamic state capable of translational (side-to-side) movement, but not "flip-flop" type movements. Hence, both the phospholipid and protein components are amphipathic, with a highly polar end in contact with the aqueous phase and hydrophobic residues buried within the membrane. Integral proteins may either be embedded in the membrane or exposed on only one side, or they may extend completely through the membrane with different portions of the proteins exposed to opposite sides of the membrane. In contrast to the fluid mosaic model, the models of protein-coated bimolecular layer of lipid diagramed in A and the unit membrane "railroad track" shown in D (which is based on osmium tetroxide fixed membranes) suggest that membranes are simply bimolecular layers of lipid coated with protein that does not penetrate the lipid. A simple bimolecular layer of lipid is shown in C, and a micelle of lipids is diagrammatically illustrated in E.

436. The answer is C. *(Stryer, 4/e, pp 268–272.)* Membrane lipids are mainly phospholipids, sphingolipids, or glycoproteins. All are constructed in the same general way. They are amphipathic molecules containing both hydrophilic and hydrophobic portions. The hydrophilic unit is a polar head group, which is often depicted as a circle. The polar head group can be phosphorylated alcohol or a sugar residue. The hydrophobic moiety is hydrocarbon tails of fatty acids. The most favored structures for phospholipids or glycolipids in aqueous media are bimolecular sheets. The formation of these lipid bilayers is a self-assembly process. Sonication of lipid bilayers will form lipid vesicles or liposomes. Lipid bilayer membranes have an extremely low permeability for ions and most polar molecules. It is the addition of proteins into the lipid bilayers that allows membranes to develop specificity and an asymmetric nature. For example, in most plasma membranes, the carbohydrate portion of glycoproteins is found only extending from the external surface of the cell.

437. The answer is D. *(Stryer, 4/e, pp 275–280.)* The fluid mosaic membrane model of S.J. Singer and G. Nicholson states that membranes are composed of globular proteins and lipids in two-dimensional solutions; that is, lipids serve as a solvent in which both they and proteins diffuse laterally unless otherwise constrained. In contrast, neither membrane phospholipids nor membrane proteins may freely move from one asymmetric side of the membrane to the other side (flip-flop). As in the earlier model of Davson and Danielli, a bimolecular layer composed of a hydrophobic interior and hydrophilic exterior is central to membrane structure.

438. The answer is E. *(Stryer, 4/e, pp 911–918.)* Protein synthesis occurs in the cytoplasm, on groups of free ribosomes called *polysomes,* and on ribosomes associated with membranes, termed the *rough endoplasmic reticulum.* However, proteins destined for secretion are only synthesized on ribosomes of the endoplasmic reticulum and are synthesized in such a manner that they end up inside the lumen of the endoplasmic reticulum. From there the secretory proteins are packaged in vesicles. The Golgi apparatus is involved in the glycosylation and packaging of macromolecules into membranes for secretion.

439. The answer is E. *(Stryer, 4/e, pp 509–510, 513–514, 569–572.)* The enzyme malate dehydrogenase is found in both the cytosol and inside the mitochondria. It is important for the reversible formation of oxaloacetate when malate is reduced and conversely the formation of malate from oxaloacetate when oxaloacetate is oxidized. In both cases, the reversible oxidation/reduction of NAD^+ occurs. In contrast, all the other enzymatic reactions described occur only in the mitochondria. The oxidative decarboxylation of pyruvate to form acetyl CoA occurs in the mitochondria and uses the last product of cytosolic glycolysis to form the first substrate of the citric acid cycle, acetyl CoA. The next step in the citric acid cycle is the formation of citrate from the condensation of oxaloacetate and *N*-acetyl CoA, eventually followed by the reduction of isocitrate to form α-ketoglutarate. The carboxylation of pyruvate to form oxaloacetate is in fact considered the first step of gluconeogenesis and occurs in the mitochondrial matrix. It is here that the mitochondrial form of malate dehydrogenase comes into use to shuttle oxaloacetate from inside the mitochondria to the cytosol. It does so by converting mitochondrial oxaloacetate to malate, which can then diffuse out into the cytoplasm via the malate carrier, where it is reformed into oxaloacetate. The remaining steps of gluconeogenesis then occur.

440. The answer is A. *(Stryer, 4/e, pp 489–490, 912–913, 917–921.)* No portion of the urea cycle occurs in the endoplasmic reticulum. The urea cycle is compartmentalized between the matrix of the mitochondria and the cytosol. The formation of ammonium ions by glutamate dehydrogenase and the incorporation of ammonia into carbamoyl phosphate followed by the synthesis of citrulline all occur in the mitochondrial matrix. The formation of argininosuccinate, arginine, and ultimately urea and ornithine occurs in the cytosol. In contrast, the synthesis of secretory and membrane proteins occurs on ribosomes attached to the endoplasmic reticulum. Glycosylation of proteins also occurs in the endoplasmic reticulum. Fatty acid elongation beyond C-16 as well as fatty acid desaturation also occurs in the membranes of the endoplasmic reticulum.

441. The answer is D. *(Stryer, 4/e, pp 404–407.)* Most eukaryotic cells have a cytoskeleton that allows them to migrate, change or maintain shape, and transport intracellular vesicles. Three filament networks make up the cytoskeleton: microtubules, microfilaments, and intermediate filaments. The contractile filaments responsible for intracellular movements and crawling migration are actin filaments known as *microfilaments.* Intermediate filaments are disulfide-rich filaments that lend mechanical stability to cells. Keratins of hair and skin are a form of intermediate filaments, as are neurofilaments, muscle desmin filaments, vimentin filaments, and lamins, components of the fibrous nuclear membrane meshwork. Microtubules are found in most cells and are long, tubelike structures built from helical assemblies of alternating α- and β-tubulin. They are important in maintaining cell shape. Microtubules also are the axonemes of cilia and flagella. *Chromatin* refers to the nucleoprotein combination of histones and DNA that forms eukaryotic chromosomes. It is not considered part of the cytoskeleton.

442. The answer is B. *(Stryer, 4/e, pp 263–272.)* Membranes are bilayers of phospholipids, sphingolipids, gangliosides, or combinations of these with proteins that give them specificity. Cholesterol is important in regulating the fluidity of membranes. Complex glycolipids such as gangliosides, which contain carbohydrates, are also important in determining the specificity in asymmetry of membranes. Nucleic acids are not structural components of cellular membranes.

443. The answer is B. *(Stryer, 4/e, pp 316–318.)* The flow of sodium ions into mammalian cells is coupled to the simultaneous entry of glucose. This concerted movement of two species is called *cotransport.* This is a symport system (as compared to an antiport system) because movement is in the same direction. The sodium ions entering in symport with glucose are pumped back out of the cell by an ATP-dependent, sodium-potassium-ATPase. Thus, ATP indirectly powers the transport. In mammals, ATP directly powers the transport of calcium and hydrogen ions, as well as the proton pumps in endocytic vesicles and the F-type H^+-ATPase of mitochondria. In practice, F-type H^+-ATPase synthesizes ATP using a proton gradient.

444. The answer is A. *(Stryer, 4/e, pp 514–517, 546, 581–582.)* Glucokinase in liver and hexokinase in other tissues are soluble cytoplasmic enzymes like all the enzymes of glycolysis. In contrast, all the other enzymes listed are part of morphologically discrete entities. Glycogen granules appear as electron-dense cytoplasmic particles in stained sections viewed with the electron microscope. The enzymes that catalyze the synthesis and degradation of glycogen (phosphorylase and synthase) are contained within the 100- to 400-Å glycogen particles, which are easily isolated by ultracentrifugation. The 300-Å diameter polyhedral structure of the three enzymes (pyruvate dehydrogenase component, dihydrolipoyl transacetylase, and dihydrolipoyl dehydrogenase) that forms the pyruvate dehydrogenase complex can be dissociated by alkaline pH followed by urea treatment. The three enzymes spontaneously reassemble into the complex at neutral pH in the absence of urea. Likewise, the large fatty acid synthetase multienzyme complex is particulate. Adenosine triphosphate (ATP) is synthesized from the proton gradient set up in mitochondrial inner membranes by an enzyme located in submitochondrial particles projecting as spheres on the matrix side of the membrane. These 85-Å knobs, which are called *coupling factor 1 (F1),* exhibit ATPase activity when mechanically isolated. The ATPase activity of isolated F1 is the reverse of its physiologic reaction, which is the synthesis of ATP.

445. The answer is B. *(Stryer, 4/e, pp 691, 927–928.)* Lysosomes are cellular organelles containing hydrolyic enzymes that degrade macromolecules taken up by these organelles. The enzymes include ribonuclease, which hydrolyzes RNA; β-galactosidase, which hydrolyzes oligosaccharides at an acid pH; acid phosphatase, which attacks phosphate esters (also at an acid pH); and cathepsins, which are intracellular proteolytic enzymes that hydrolyze peptide bonds. Congenital lack of certain lysosomal enzymes can lead to accumulation of the normally degraded substrate. For example, lack of *N*-acetylhexosaminidase leads to accumulation of ganglioside G_{M2} in the deadly Tay-Sachs disease. Lipoprotein lipase is an enzyme released into the bloodstream from capillary endothelial cells during delipidation of blood lipoproteins.

446–448. The answers are 446-C, 447-E, 448-B. *(Stryer, 4/e, pp 351–352, 612–613, 622–623.)* The designated sites labeled in the electron micrograph of a cell are **A**, nucleus; **B**, rough endoplasmic reticulum; **C**, mitochondria; and **D**, lipid droplets. The ketone bodies acetoacetate and β-hydroxybutyrate are synthesized in the liver, released into the bloodstream, and taken up and utilized as a source of acetate by peripheral tissues during starvation or fasting. In order to be activated to acetyl CoA, acetoacetate is converted to acetoacetyl CoA by succinyl CoA transferase, which utilizes succinyl CoA, a normal metabolite of the mitochondrial citric acid cycle.

The insulin receptor is found on the plasma membranes of cells. It is activated by the binding of insulin and turns on transport of glucose and amino acids by an as-yet-unknown mechanism.

Desaturation of fatty acids occurs in the endoplasmic reticulum. Mixed-function oxidases insert double bonds into fatty acids by a process utilizing molecular oxygen and NADPH.

Metabolism

DIRECTIONS: Each question below contains five suggested responses. Select the **one best** response to each question.

449. Hormone receptors that stimulate cyclic AMP production

(A) are part of a complex of two proteins that transform the external signal into internal cyclic AMP production
(B) cause release of the catalytic subunit upon binding of the hormone
(C) are proteins distinct and separate from those that catalyze the production of cyclic AMP
(D) are the outer protruding portion of a single protein that acts as a receptor and an enzyme that produces cyclic AMP
(E) are not very specific and bind a number of different hormones

450. The oxygen dissociation curve for hemoglobin is shifted to the right by

(A) decreased O_2 tension
(B) decreased CO_2 tension
(C) increased CO_2 tension
(D) increased N_2 tension
(E) increased pH

451. A 0.22 *M* solution of lactic acid (pK_a 3.9) was found to contain 0.20 *M* in the dissociated form, and 0.02 *M* undissociated. What is the pH of the solution?

(A) pH 2.9
(B) pH 3.3
(C) pH 3.9
(D) pH 4.9
(E) pH 5.4

452. Which of the lettered points on the graph below best represents completely compensated metabolic acidosis?

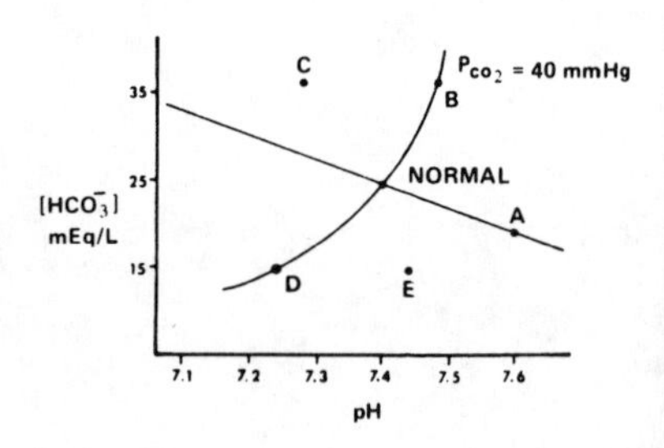

(A) A
(B) B
(C) C
(D) D
(E) E

453. The component in muscle tissue that contains the adenosine triphosphatase (ATPase) activity required for contraction is

(A) actin
(B) myosin
(C) sarcoplasmic reticulum
(D) motor end plate
(E) calcium

454. A healthy 70-kg man eats a well-balanced diet containing adequate calories and 62.5 g of high-quality protein per day. Measured in grams of nitrogen, his daily nitrogen balance would be

(A) + 10 g
(B) +6.25 g
(C) 0 g
(D) −6.25 g
(E) −10 g

455. The antimalarial drug primaquine may induce hemolytic anemia via a deficiency in

(A) glucose-6-phosphatase
(B) glucose-6-phosphate dehydrogenase
(C) glucose-6-phosphate isomerase
(D) glucokinase
(E) hexokinase

456. The pH of body fluids is stabilized by buffer systems. Which of the following compounds is the most effective buffer at physiologic pH?

(A) Na_2HPO_4, $pK_a = 12.32$
(B) NH_4OH, $pK_a = 9.24$
(C) NaH_2PO_4, $pK_a = 7.21$
(D) CH_3CO_2H, $pK_a = 4.74$
(E) Citric acid, $pK_a = 3.09$

457. Water, which constitutes 70 percent of body weight, may be said to be the "cell solvent." The property of water that most contributes to its ability to dissolve compounds is the

(A) strong covalent bond formed between water and salts
(B) hydrogen bond formed between water and biochemical molecules
(C) hydrophobic bond formed between water and long-chain fatty acids
(D) absence of interacting forces
(E) fact that the freezing point of water is much lower than body temperature

458. In patients fed an equal amount (on a molar basis) of carbohydrates and fats, the respiratory quotient is

(A) 2.72
(B) 1.00
(C) 0.86
(D) 0.72
(E) 0.10

459. The major source of extracellular cholesterol for human tissues is

(A) very low-density lipoprotein (VLDL)
(B) low-density lipoprotein (LDL)
(C) high-density lipoprotein (HDL)
(D) albumin
(E) γ-globulin

460. Which of the schematic configurations below represents a stable lipid-water interaction?

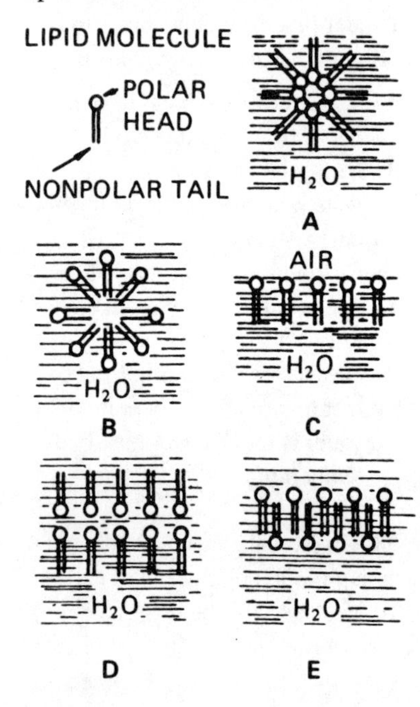

(A) Figure A
(B) Figure B
(C) Figure C
(D) Figure D
(E) Figure E

461. In the normal resting state of humans, most of the blood glucose burned as fuel is consumed by

(A) the liver
(B) the brain
(C) the kidneys
(D) adipose tissue
(E) muscle

462. Most major metabolic pathways are considered to be either mainly anabolic or catabolic. Which of the following pathways is most correctly considered to be amphibolic?

(A) Lipolysis
(B) Glycolysis
(C) β-Oxidation of fatty acids
(D) Citric acid cycle
(E) Gluconeogenesis

463. McArdle's syndrome (a glycogen storage disease) involves a deficiency of which of the following enzymes?

(A) Pancreatic peptidase
(B) Hepatic glycogen synthetase
(C) Hepatic phosphorylase
(D) Muscle phosphorylase
(E) Debranching enzyme

464. The synthesis of 3-hydroxy-3-methylglutaryl CoA can occur

(A) only in mitochondria of all mammalian tissues
(B) only in the cytosol of all mammalian tissues
(C) only in the endoplasmic reticulum of all mammalian tissues
(D) in both the cytosol and mitochondria
(E) in lysosomes

465. Phenylketonuria (PKU) is thought to be a genetic disease caused by

(A) thiamine processing deficiency
(B) deficiency of a phenylalanine transport enzyme
(C) tyrosinase deficiency
(D) deficiency of a hydroxylase involved in an amino acid conversion
(E) an excess of tyrosine

466. Which one of the following amino acids gives rise to α-ketoacids that accumulate in the urine in maple syrup urine disease?

(A) Lysine
(B) Arginine
(C) Glutamate
(D) Phenylalanine
(E) Valine

467. Which one of the following products of protein metabolism is decreased below normal levels during the early stages of starvation?

(A) Urea
(B) Anabolic enzymes
(C) CO_2
(D) NH_4^+
(E) Glucose

468. Elevation of cytosolic calcium from intracellular stores for smooth muscle contraction can be caused by

(A) cyclic AMP
(B) cyclic GMP
(C) cyclic CMP
(D) inositol 1,4,5-triphosphate (IP_3)
(E) diacylglycerol (DAG)

469. During synthesis of mature collagen fiber, which one of the following steps would occur within the fibroblast?

(A) Hydrolysis of procollagen to form collagen
(B) Glycosylation of proline residues
(C) Formation of a triple helix
(D) Formation of covalent cross-links between molecules
(E) Assembly of the collagen fiber

470. When the liver is actively synthesizing fatty acids, a concomitant decrease in β-oxidation of fatty acids is due to

(A) inhibition of a translocation between cellular compartments
(B) inhibition by an end product
(C) activation of an enzyme
(D) detergent effects
(E) decreases in adipocyte lipolysis

471. Glycolysis is the only ATP-producing pathway in

(A) erythrocytes
(B) lymphocytes
(C) hepatocytes
(D) neurons
(E) adipocytes

472. Which one of the following is a correct statement about the regulation and sequence of reactions in metabolic pathways?

(A) The initial step in many pathways is a major determinant of control
(B) The sequence of steps in catabolic pathways is usually the exact reverse of the biosynthetic sequence
(C) Enzymes found in an anabolic pathway are rarely found in the corresponding catabolic pathway
(D) A small set of large precursors serves as the starting point for most biosynthetic processes in energy metabolism
(E) Steps in both anabolic and catabolic pathways are usually irreversible

473. Which one of the following tissues can metabolize glucose, fatty acids, and ketone bodies for ATP production?

(A) Liver
(B) Muscle
(C) Hepatocytes
(D) Brain
(E) Red blood cells

474. The problem of regenerating NAD^+ from NADH for cytoplasmic processes by using mitochondria is solved in the most energy-efficient manner by which one of the following intracelluar shuttle systems?

(A) Citrate → pyruvate shuttle
(B) Dihydroxyacetone phosphate → α-glycerophosphate shuttle
(C) Malate → aspartate shuttle
(D) Citrate → citrate shuttle
(E) Lactate → pyruvate shuttle

475. Which one of the following statements is correct regarding the well-fed state?

(A) NADPH production by the hexose monophosphate shunt is decreased
(B) Acetoacetate is the major fuel for muscle
(C) Glucose transport into adipose tissue is decreased
(D) The major fuel used by the brain is glucose
(E) Amino acids are utilized for glucose production

476. Which one of the following is a true observation of obese patients with non-insulin-dependent diabetes mellitus?

(A) They show marked improvement in glucose tolerance tests when body weight is reduced to normal
(B) A course of insulin approximately 4 h after each meal is warranted
(C) They are difficult to diagnose since they show a normal glucose tolerance test
(D) They demonstrate lower plasma levels of glucagon compared with normal persons
(E) They have extraordinarily low levels of insulin in the blood compared with normal persons

477. During an overnight fast, the major source of blood glucose is

(A) dietary glucose from the intestine
(B) hepatic glycogenolysis
(C) gluconeogenesis
(D) muscle glycogenolysis
(E) glycerol from lipolysis

478. The jinga bean, found in the jungles of Brazil, is unique in that it is composed almost exclusively of protein. Studies have shown that immediately following a meal composed exclusively of jinga beans, which one of the following occurs?

(A) A decreased release of epinephrine
(B) A complete absence of liver glycogen
(C) Hypoglycemia
(D) An increased release of insulin
(E) Ketosis caused by the metabolism of ketogenic amino acids

DIRECTIONS: Each numbered question or incomplete statement below is NEGATIVELY phrased. Select the **one best** lettered response.

479. Approximately 3 h following a well-balanced meal, blood levels of all the following would be expected to be elevated EXCEPT

(A) fatty acids
(B) insulin
(C) very low-density lipoproteins
(D) glucose
(E) chylomicrons

480. The absorption of glucose from the gut into intestinal mucosal cells is coupled to Na^+,K^+-ATPase. In contrast, the movement of glucose from the intestinal epithelial cells into the submucosal bloodstream occurs through passive transport. Given these facts, all the following statements can be true at one time or another EXCEPT

(A) cytosolic levels of glucose in intestinal mucosal cells are unaffected by levels of glucose in skeletal muscle cells
(B) free glucose levels in the lumen of the intestine can be higher than intestinal cell levels
(C) plasma glucose levels are much higher than intestinal cell cytosolic levels of glucose
(D) levels of glucose in the intestinal lumen can be lower than those found in the cytosol of intestinal epithelial cells
(E) the levels of plasma glucose are approximately equal to those of the cytosol of intestinal epithelial cells

481. Pathways common to many tissues include all the following EXCEPT

(A) glycolysis
(B) lipolysis
(C) ketogenesis
(D) glycogenolysis
(E) hexose monophosphate pathway

482. A homogenate of liver cells is centrifuged at 100,000 × g for 1 h. Following this, the supernatant is separated from the pellet and the pellet is resuspended in physiologic medium. Assuming inclusion of the appropriate substrates and cofactors, all the following enzymatic activities can be measured in the resuspended pellet EXCEPT

(A) glucose-6-phosphate dehydrogenase
(B) glycogen synthetase
(C) aconitase
(D) acyl CoA hydratase
(E) hydroxybutyrate dehydrogenase

483. All the following descriptions of calcium are correct EXCEPT

(A) it diffuses as a divalent cation
(B) it is required as a cofactor for many reactions
(C) it freely diffuses across the endoplasmic reticulum of muscle cells
(D) it is most highly concentrated in bone
(E) it is excreted through the intestinal lumen

484. All the following enzymes or events play a major role in adipocytes EXCEPT

(A) lipolysis
(B) glycerol kinase
(C) hormone-sensitive triacylglyceride lipase
(D) glycolysis
(E) phosphatidate phosphatase

485. Deficiencies of all the following enzymes result in enlarged livers and hypoglycemia EXCEPT

(A) amylo-(1,4→1,6)-transglycosylase
(B) phosphorylase
(C) phosphorylase kinase
(D) amylo-1,6-glucosidase
(E) glucose-6-phosphatase

486. All the following may be derived in whole or in part from amino acids in humans EXCEPT

(A) sphingosine
(B) acetyl CoA
(C) pantothenic acid
(D) glucose
(E) niacin

487. All the following tissues are capable of contributing to blood glucose EXCEPT

(A) liver
(B) kidneys
(C) cardiac muscle
(D) duodenal epithelium
(E) jejunal epithelium

488. All the following are membranous elements of cardiac or skeletal muscles EXCEPT

(A) endoplasmic reticulum
(B) sarcolemma
(C) intercalated disks
(D) sarcoplasmic reticulum
(E) sarcomere

DIRECTIONS: Each group of questions below consists of lettered headings followed by a set of numbered items. For each numbered item select the **one** lettered heading with which it is **most** closely associated. Each lettered heading may be used **once, more than once, or not at all.**

Questions 489–493

Match each description with the correct disease.

(A) McArdle's phosphorylase disease
(B) Diabetes
(C) Carnitine deficiency
(D) Thalassemia
(E) Hyperammonemia
(F) Hypercholesterolemia
(G) Methanol poisoning
(H) Cancer
(I) Lesch-Nyhan syndrome
(J) Hemophilia
(K) Xeroderma pigmentosum
(L) Diphtheria

489. Loss of ability to move specific molecules between membrane-separated cellular compartments

490. Cessation of protein synthesis

491. Treatment by competitive inhibition

492. Impaired RNA synthesis or processing

493. Deficiency of DNA repair

Questions 494–496

Match each characteristic with the correct enzyme.

(A) Lipoprotein lipase
(B) Fructose bisphosphate phosphatase I
(C) Fructose-2,6-bisphosphate
(D) Pancreatic lipase
(E) Phospholipase A_2
(F) Phosphofructokinase I
(G) Hormone-sensitive lipase

494. Activated by dephosphorylation

495. Acts on plasma lipoproteins

496. Allosterically activated

Questions 497–500

For each patient below, select the set of concentrations of fuels in the blood that most closely correlates with that patient.

	Concentration of Blood Fuels (mM)			
	Glucose	Free Fatty Acids	Ketone Bodies	Amino Acids
(A)	2.00	3.0	10.00	5.0
(B)	4.50	1.5	5.00	4.7
(C)	12.00	2.0	10.00	4.5
(D)	4.50	0.5	0.02	4.5
(E)	4.49	2.0	8.00	3.1

497. A normal patient

498. A patient following 4 days of fasting

499. A patient following 1 month of starvation

500. A patient with uncontrolled diabetes mellitus

Metabolism

Answers

449. The answer is C. *(Stryer, 4/e, pp 340–343.)* To date, at least three separate and distinct proteins have been implicated in the translation of hormone action to intracellular cyclic AMP production. These include the very specific hormone receptor, an intermediate guanyl nucleotide-binding protein (G protein), and the catalytic unit that converts ATP into cyclic AMP. The guanyl nucleotide-binding protein binds the hormone-occupied receptor to the catalytic unit of adenylate cyclase. When stimulated by a hormone-occupied receptor, the inactive G protein exchanges GDP for GTP and thereby becomes activated. In the absence of hormone-receptor complex, GTP is hydrolyzed to GDP and the G protein becomes inactivated.

450. The answer is C. *(Stryer, 4/e, pp 159–160.)* The curve representing the binding of oxygen to hemoglobin as a function of the partial pressure of oxygen is sigmoidal in shape. A shift to the right indicates a decrease in the capacity of hemoglobin to bind oxygen. Lowering the pH reduces the affinity of hemoglobin for oxygen, as does increasing the partial pressure of CO_2 at constant pH. The increased concentrations of H^+ and CO_2 therefore encourage the release of oxygen by hemoglobin. This is known as the *Bohr effect.*

451. The answer is D. *(Rawn, pp 40–44. Stryer, 4/e, pp 42–43.)* According to the Henderson-Hasselbalch equation, $pH = pK_a + \log \frac{\text{base}}{\text{acid}}$. In the case of 0.2 *M* lactate and 0.02 *M* lactic acid as presented in the question, pH = 3.9 + log 10 = 4.9.

452. The answer is E. *(Stryer, 4/e, pp 42–43, 164–165.)* In metabolic acidosis, blood bicarbonate is found to be low. In a situation of metabolic acidosis where bicarbonate has been consumed, compensation can be made by the respiratory mechanism of hyperventilation (sometimes referred to as "blowing off CO_2"), whereby pH is restored to normal by reducing plasma carbon dioxide.

453. The answer is B. *(Stryer, 4/e, pp 393–394.)* The adenosine triphosphatase (ATPase) that hydrolyzes adenosine triphosphate (ATP), thereby releasing the energy that allows for muscle contraction, is located on the myosin molecule. Myosin binds to the polymerized form of actin. This interaction, coupled with the ATPase activity of myosin, is critical for the generation of the force that translates myosin and actin relative to each other during contraction. The other components of muscle listed in the question lack ATPase activity.

454. The answer is C. *(Stryer, 4/e, pp 634–636.)* The daily intake of 62.5 g of high-quality protein is above the minimum daily requirement for a 70-kg adult (45 g protein per day). As the obligatory nitrogen losses are covered by the dietary intake, this man will be in nitrogen balance (i.e., 0) and nitrogen loss will equal nitrogen intake.

455. The answer is B. *(Stryer, 4/e, pp 567–568.)* In approximately 11 percent of black Americans, a tenfold reduction in the activity of glucose-6-phosphate dehydrogenase in red blood cells is observed. This seems to be a protective genetic mechanism against falciparum malaria. A decrease in the activity of glucose-6-phosphate dehydrogenase is characteristic of primaquine-sensitive erythrocytes, and when patients with this deficiency are given the antimalarial drug, severe symptoms are noted. Black urine, jaundice, and a decrease in hemoglobin are observed and are indicative of massive and possibly fatal destruction of red blood cells. Glucose-6-phosphate dehydrogenase represents the first enzymatic step in the pentose phosphate pathway, which is the only source of NADPH in red cells, which lack mitochondria. A decreased amount of NADPH slows down the function of glutathione reductase. The latter enzyme reduces the oxidized disulfide form of glutathione to the sulfhydryl form. Reduced glutathione maintains the structure of red blood cells by keeping hemoglobin in the ferrous state and other red cell proteins in the reduced state. Thus primaquine may induce hemolysis in patients who already have a reduced level of glucose-6-phosphate dehydrogenase.

456. The answer is C. *(Stryer, 4/e, pp 42–43.)* In any fluid, the maximum buffering action is achieved by the acid whose pK_a most nearly approximates the pH of the fluid. Physiologic pH is about 7.4, so that among those buffers listed in the question, NaH_2PO_4 is the most effective.

457. The answer is B. *(Stryer, 4/e, pp 9–11.)* Water molecules have a dipole nature and dissolve salts because of attractions between the water dipoles and the ions that exceed the force of attraction between the oppositely charged ions of the salt. In addition, the latter force is weakened by the high dielectric constant of water. Nonionic, but polar, compounds are dissolved in water because of hydrogen bonding between water molecules and groups such as alcohols, aldehydes, and ketones.

458. The answer is C. *(Stryer, 4/e, pp 551–552.)* The respiratory quotient (RQ) can be studied noninvasively by measuring the CO_2 produced and the O_2 consumed following a meal.

$$RQ = \frac{\text{moles of } CO_2 \text{ produced}}{\text{moles of } O_2 \text{ consumed}}$$

The RQ varies for each major food group. The RQ is 0.8 for proteins, 0.72 for fats, and 1.0 for carbohydrates. Thus, if more or less equal amounts of fats and carbohydrates are ingested and burned for energy, the RQ will be about halfway between the values of either foodstuff alone, i.e., 0.86.

459. The answer is B. *(Stryer, 4/e, pp 697–701.)* The uptake of exogenous cholesterol by cells results in a marked suppression of endogenous cholesterol synthesis. Low-density human lipoprotein not only contains the greatest ratio of bound cholesterol to protein but also has the greatest potency in suppressing endogenous cholesterogenesis.

460. The answer is B. *(Stryer, 4/e, pp 268–270.)* Lipid micelles are stable in water when the polar heads of the lipid face outward in contact with water and the hydrophobic, nonpolar tails turn inward to exclude water (figure **B**). None of the other configurations shown in the question represent a thermodynamically favorable interaction between the hydrophobic groups and water.

461. The answer is B. *(Stryer, 4/e, pp 770–773.)* Although the brain accounts for only about 20 percent of the calories burned by an average human during a normal day, it consumes up to 120 g of glucose per day. This corresponds to approximately 420 kcal and amounts to about 60 percent of the total utilization of glucose by the whole body. This is due to the fact that the brain requires glucose as a virtually exclusive energy source. However, during prolonged starvation, ketone bodies can replace glucose as the major source of fuel for the brain. The preferred fuels of resting muscle are ketone bodies and fatty acids, whereas both the liver and kidneys can utilize a variety of fuels. Although glucose-derived formation of α-glycerolphosphate is the major controlling factor in esterification of triacylglycerols for storage in adipose cells, these cells can utilize both glucose and fatty acids as fuels to generate energy.

462. The answer is D. *(Stryer, 4/e, pp 766–769.)* In general, the corresponding pathways of catabolism and anabolism are not identical (glycolysis versus gluconeogenesis, lipolysis and β-oxidation of fatty acids versus fatty acid synthesis and lipogenesis, glycogenolysis versus glycogenesis).

However, the citric acid cycle is a central pathway from which anabolic precursors of biosynthetic reactions may derive or into which the complete catabolism of small molecules to carbon dioxide and water may occur. For these reasons, the citric acid cycle is often called an *amphibolic* pathway.

463. The answer is D. *(Stryer, 4/e, pp 598–599.)* Muscle phosphorylase deficiency leads to a glycogen storage disease (McArdle's syndrome) and, in young adults, an inability to do strenuous physical work because of muscular cramps resulting from ischemia. The compromised phosphorylation of muscle glycogen characteristic of McArdle's syndrome compels the muscles to rely on auxiliary energy sources such as free fatty acids and ambient glucose.

464. The answer is D. *(Stryer, 4/e, pp 612–613, 693.)* The synthesis of 3-hydroxy-3-methylglutaryl CoA requires the condensation of three acetyl CoA groups. The two enzymatic steps involved are the first two steps of cholesterol synthesis and ketone body synthesis. While cholesterol synthesis occurs in the cytosol of most mammalian tissues, ketone body synthesis can only occur in the mitochondria of liver cells. Not only are cholesterol synthesis and ketone body synthesis separated by compartmentalization, they are separated by metabolic needs. Cholesterol synthesis is an anabolic pathway that takes place when acetyl CoA production from excess dietary precursors is possible. In contrast, ketone body production by the liver occurs when acetyl CoA levels from β-oxidation are high. This catabolic situation exists during fasting, starvation, and uncontrolled diabetes.

465. The answer is D. *(Stryer, 4/e, pp 649–650.)* Phenylketonuria is thought to result from a genetic deficiency in the enzyme phenylalanine hydroxylase, which converts phenylalanine to tyrosine. This disorder, which results in dementia, seizures, and dermatologic manifestations, is largely preventable by adequate perinatal screening and treatment. The treatment consists of a special diet low in phenylalanine.

466. The answer is E. *(Stryer, 4/e, pp 645–646.)* Maple syrup urine disease results from a deficiency of the α-ketoacid dehydrogenases responsible for oxidation of the α-keto analogues of the branched-chain amino acids (leucine, isoleucine, and valine). Consequently, these α-ketoacids accumulate in blood, urine, and spinal fluid and produce the biochemical abnormalities and odor characteristic of this disorder. The disease is transmitted as an autosomal recessive trait and clinical features include mental and physical retardation.

467. The answer is B. *(Stryer, 4/e, pp 775–777.)* During the early phases of starvation, the catabolism of proteins is at its highest level. Anabolic enzymes, which are not utilized during starvation, are degraded and their synthesis is repressed. The deamination of amino acids for gluconeogenesis and ketogenesis results in a negative nitrogen balance. Hence, ammonia and urea levels in the urine exceed normal values. The glucose formed from gluconeogenic amino acids becomes the major source of blood glucose following depletion of liver glycogen stores. Complete oxidation of this glucose, as well as the ketone bodies formed from ketogenic amino acids, leads to a relative increase in the CO_2 and H_2O formed from amino acid carbon skeletons.

468. The answer is D. *(Stryer, 4/e, pp 343–345.)* Inositol 1,4,5-triphosphate (IP_3) and diacylglycerol (DAG) are the end result of the phosphoinositide cascade. This cascade, like the cyclic AMP cascade, forms intracellular second messengers in response to extracellular signals. IP_3 causes the release of calcium into the cytosol of cells, while DAG activates protein kinase C. In the case of smooth muscle cells, this results in contraction.

469. The answer is C. *(Stryer, 4/e, pp 31–32.)* The connective tissue fiber collagen is synthesized by fibroblasts. However, because the length of the finished collagen fibers is many times greater than that of the cell of its origin, a portion of its assembly occurs extracellularly. The intracellular formation of the biosynthetic precursor of collagen, procollagen peptides pro-α1(I) and pro-α2, occurs in the following steps: (1) polypeptide synthesis, (2) hydroxylation of proline and lysine residues, (3) glycosylation of lysine residues (proline residues are not glycosylated), (4) triple-helix formation, and (5) secretion. Once outside the fibroblasts, procollagen molecules are activated by fibroblast-specific procollagen peptidases. Before specific proteolytic cleavage of procollagen, tropocollagen bundles will not assemble into collagen fibers. Once the collagen fibers are formed, aldo cross-links between lysine residues and histidine-aldo cross-links are formed. These cross-links covalently bind the collagen chains to one another. The extent and type of cross-linking determines the flexibility and strength of the collagen mass formed.

470. The answer is A. *(Stryer, 4/e, pp 621–622.)* Under conditions of active synthesis of fatty acids in the cytosol of hepatocytes, levels of malonyl CoA are high. Malonyl CoA is the activated source of two carbon units for fatty acid synthesis. Malonyl CoA inhibits carnitine acyltransferase I, which is located on the cytosolic face of the inner mitochondrial membrane. Thus, long-chain fatty acyl CoA units cannot be transported into mitochondria where β-oxidation occurs, and translocation from cytosol to mitochondrial matrix is prevented. In this situation compartmentalization of membranes as well as inhibition of enzymes comes into play.

471. The answer is A. *(Stryer, 4/e, pp 483–484, 509–510.)* While erythrocytes can produce ATP by metabolizing glucose anaerobically to lactate via the glycolytic pathway, this process does not involve any net oxidation of glucose. Erythrocytes have no mitochondria for oxidative processes. The reduced form of nicotinamide adenine dinucleotide (NADH) derived from the triose phosphate dehydrogenase step is used to reduce pyruvate to lactate. All other tissues with mitochondria are widely adaptable in their metabolic requirements and can oxidize glucose to produce ATP.

472. The answer is A. *(Stryer, 4/e, pp 259–260.)* Although the same intermediates may appear in both anabolic and catabolic pathways, one path is not simply the reverse of the other because irreversible enzymatic steps often occur in the beginning of the reaction sequence. However, many steps in both anabolic and catabolic pathways are reversible. The same enzymes often appear in many metabolic pathways, but regulatory steps are irreversible. A number of small precursors serve as the building blocks of anabolism, while large energy-storage molecules such as glycogen, lipids, and proteins give rise to smaller molecules during catabolic processes.

473. The answer is B. *(Stryer, 4/e, pp 569–576.)* Muscle cells are the only cells listed that are capable of utilizing all the energy sources available—glucose, fatty acids, and, during fasting, ketone bodies. Mitochondria are required for metabolism of fatty acids and ketone bodies. Since red blood cells (erythrocytes) do not contain mitochondria, no utilization of these energy sources is possible. Although brain may utilize glucose and ketone bodies, fatty acids cannot cross the blood-brain barrier. Hepatocytes, or liver cells, are the sites of ketone body production, but the mitochondrial enzyme necessary for utilization of ketone bodies is not present.

474. The answer is C. *(Stryer, 4/e, pp 548–549.)* NADH generated from glycolysis must be relieved of an electron to form nicotinamide adenine dinucleotide (NAD^+) so that glycolysis may continue. However, mitochondrial membranes are impermeable to both NADH and NAD^+. The solution to this problem is the transfer of electrons from NADH to molecules that will traverse the membrane. In the glycerophosphate shuttle, dihydroxyacetone phosphate (DHAP) is reduced to glycerol-3-phosphate and thereby regenerates NAD^+. The glycerol-3-phosphate diffuses into mitochondria and is oxidized by flavin adenine dinucleotide (FAD) back to DHAP, which can diffuse back into the cytosol. The mitochondrial reduced form of flavin adenine dinucleotide ($FADH_2$) produced yields 2 ATP in the electron transport chain. In heart and liver, the more energy-efficient malate-aspartate shuttle moves electrons into mitochondria. Cytoplasmic oxaloacetate is reduced to malate, which diffuses into the mitochondria and is oxidized by NAD^+ back to oxaloacetate. The mitochondrial NADH produced yields 3 ATP on electron transport. The mitochondrial oxaloacetate is converted to aspartate, which diffuses into the cytosol, where it is converted back into cytoplasmic oxaloacetate.

475. The answer is D. *(Stryer, 4/e, pp 775–777.)* Glucose is the major fuel for the brain in the well-fed state. The brain requires a continuous supply of glucose at all times. In fact, if glucose drops to a low level, convulsions may follow. However, during starvation or fasting, the brain is capable of obtaining approximately 75 percent of its energy from circulating ketone bodies. During the absorptive phase, ketone bodies such as acetoacetate and 3-hydroxybutyrate are low. Circulating amino acids are utilized for protein synthesis. Liver production of NADPH is at a high level since it is needed for fatty acid synthesis. Glucose is actively transported into all cells, including adipocytes, which require it to form glucose-3-phosphate for esterifying fatty acids into triacylglyceride.

476. The answer is A. *(Stryer, 4/e, pp 774–780.)* The vast majority of patients with non-insulin-dependent diabetes mellitus are overweight. Virtually all of them show improvement in blood glucose tolerance tests with weight reduction to normal levels. In the obese state, all of these patients show an abnormal glucose tolerance test. Most non-insulin-dependent diabetics have elevated insulin levels. They do not derive any benefit from added insulin since their problem is not a lack of insulin but a lack of response to insulin. Glucagon levels are usually in the normal range. For any patient, the injection of insulin 4 h after a meal might well have hypoglycemic effects since glucose levels may be down by this time.

477. The answer is B. *(Stryer, 4/e, pp 773–777.)* In the absorptive phase following a meal, the major source of glucose is glucose taken directly from the intestine into the blood system. Much of this glucose is absorbed into cells and, in particular, into the liver via the action of insulin, where it is stored as glycogen. Once the effects of daytime eating have subsided and all the glucose from absorption has been stored, the normal overnight fast begins. During this period, the major source of blood glucose is hepatic glycogen. Through the effects of glycogenolysis, which are mediated by glucagon, hepatic glycogen is slowly parceled out as glucose to the bloodstream, keeping blood glucose levels normal. In contrast, muscle glycogenolysis has no effect on blood glucose levels because no glucose-6-phosphatase exists in muscle and hence phosphorylated glucose cannot be released from muscle into the bloodstream. Following a more prolonged fast or the early stages of starvation, gluconeogenesis is needed to produce glucose from glucogenic amino acids and the glycerol released by lipolysis of triacylglycerides in adipocytes. This is because the liver glycogen is depleted and the liver is forced to turn to gluconeogenesis to produce the amounts of blood glucose necessary to maintain blood levels.

478. The answer is A. *(Stryer, 4/e, pp 770–777.)* High blood levels of amino acids, in addition to glucose, promote the release of insulin through their action on receptors at the surface of the beta cells of the pancreas. While insulin alone could lead to a hypoglycemic effect, hypoglycemia should not be observed since glucagon is also released in response to the elevated levels of circulating amino acids. The balance of glucagon and glucose tends to keep blood levels of glucose within normal ranges while amino acid transport into cells is promoted. Due to the normal insulin levels for a fed state, ketosis and depletion of liver glycogen are not observed. Both of these events occur during fasting and starvation due to the abundance of glucagon and epinephrine in the blood as opposed to the low levels of insulin.

479. The answer is A. *(Stryer, 4/e, pp 774–777.)* Following digestion, the products of digestion enter the bloodstream. These include glucose, amino acids, triacylglycerides packaged into chylomicrons from the intestine, and very low-density lipoproteins from the liver. The hormone of anabolism, insulin, is also elevated because of the signaling of the glucose and amino acids in the blood, which allows its release from the beta cells of the pancreas. Insulin aids the movement of glucose and amino acids into cells. In contrast, all the hormones and energy sources associated with catabolism are decreased in the blood during this time. Long-chain fatty acids and glycerol released by lipolysis from adipocytes are not elevated. Glucose levels can be elevated; in fact, the only time that glucose levels rise significantly above approximately 80 mM is following a well-balanced meal when glucose is obtained from the diet.

480. The answer is C. *(Stryer, 4/e, pp 316–318.)* The plasma membranes of intestinal epithelial cells contain a sodium gradient that drives the active transport of glucose. The rate and amount of glucose transported depend upon the sodium gradient maintained across the plasma membrane. Sodium ions entering the cell in the company of glucose are pumped out again by Na^+,K^+-ATPase. Once in the cytosol of the intestinal cell, the glucose moves across the cell and diffuses out of the cell into the interstitial fluid of the submucosa and then into the plasma of the capillaries underlying the intestinal epithelium. This occurs for the following reason: While glucose is maintained in blood plasma at an approximately constant level, it is always slowly moving out of the plasma into the cells of tissue that use it. Given that the diffusion from the intestinal cells into the plasma is passive, the intestinal cells and the plasma try to maintain an equilibrium. Thus, plasma glucose levels are always approximately equal to or slightly less than those found in the intestinal cells. Due to the passive maintenance of this equilibrium, it is highly unlikely that the concentration of glucose in the plasma can get much higher than that in the intestinal cell cytosol. It is also unlikely that the levels of glucose in other tissues of the body (for example, muscle) will have any bearing upon those found in the intestinal cell.

481. The answer is C. *(Stryer, 4/e, pp 766–768.)* The release of glucose from glycogen (glycogenolysis), the breakdown of glucose to pyruvate (glycolysis), the release of fatty acids and glycerol from triacylglycerol stores (lipolysis), and the production of NADPH and other sugars from glucose-6-phosphate (hexose monophosphate pathway) are pathways found in many tissues. In contrast, ketogenesis, which is the formation of 3-hydroxybutyrate and acetoacetate from acetyl CoA, occurs only in liver mitochondria. Utilization of ketone bodies as an energy source is carried out by many tissues but not by liver.

482. The answer is A. *(Stryer, 4/e, pp 560, 766–768.)* Centrifugation of a cellular homogenate at a force of 100,000 × g will pellet all cellular organelles and membranes. Only soluble cellular molecules found in the cytosol will remain in the supernatant. Thus, the enzymes of glycolysis, and most of those of gluconeogenesis, fatty acid synthesis, and the pentose phosphate pathway will be in the supernatant. On the other hand, in the pellet will be the enzymes of the citric acid cycle (aconitase) with mitochondria; glycogen degradation and synthesis (glycogen synthetase) with glycogen particles; α-oxidation (acyl CoA hydratase) with mitochondria; and ketogenesis (hydroxybutyrate dehydrogenase) with mitochondria. Glucose-6-phosphate dehydrogenase, which results in the formation of 6-phosphoglucono-δ-lactone from glucose-6-phosphate, is the committed step in the pentose phosphate pathway. The pentose phosphate pathway occurs free in the cytosol.

483. The answer is C. *(Stryer, 4/e, pp 314–316, 344, 347–348, 402–404.)* Calcium ions and calcium deposits are virtually universal in the structure and function of living things. In humans, calcium ions are required for the activity of many enzymes. Calcium is taken up from the gut in the presence of forms of vitamin D, such as cholecalciferol. Calcium is also primarily excreted through the intestine. When soluble, it is present as a divalent cation. When insoluble, it is found as hydroxyapatite in bone. It is required by muscle cells for contraction and is sequestered into sarcoplasmic reticulum during relaxation. It is actively transported by a calcium-ATPase across the sarcoplasmic reticulum.

484. The answer is B. *(Stryer, 4/e, pp 605–606.)* The function of adipose tissue is the storage of fatty acids as triacylglycerols in times of plenty and the release of fatty acids during a time of fasting or starvation. Fatty acids taken in by adipocytes are stored by esterification to glycerol-3-phosphate. Glycerol-3-phosphate is derived almost entirely from the glycolytic intermediate dihydroxyacetone phosphate through the action of glycerol-3-phosphate dehydrogenase. Glycerol kinase is not present to any great extent in adipocytes, so that glycerol freed during lipolysis is not used to reesterify the fatty acids being released. The enzyme phosphatidate phosphatase converts phosphatidic acid to diacylglycerol during synthesis of triacylglycerides. The enzyme triacylglyceride lipase is turned on by phosphorylation by a cyclic AMP–dependent protein kinase following epinephrine stimulation.

485. The answer is A. *(Stryer, 4/e, pp 598–599.)* Amylo-(1,4 → 1,6)-transglycosylase (also known as the *branching enzyme*) functions in the separate pathway for glycogen synthesis. Lack of glycogen storage would result from its deficiency. Deficiencies in enzymes needed to break down stored liver glycogen lead to enlarged livers and hypoglycemia. Mobilization of glycogen stores to produce glucose in the liver requires the phosphorolysis of the glycogen chain by the enzyme phosphorylase, which is activated by phosphorylation catalyzed by phosphorylase kinase. Also needed are the hydrolysis of α-1,6-glycosidic bonds by amylo-1,6-glucosidase (also known as the *debranching enzyme*) and the hydrolysis of glucose-6-phosphate derived from glucose-1-phosphate (product of phosphorylase) by glucose-6-phosphatase to produce glucose for export into the blood. A genetic disease resulting in the deficiency of any of these enzymes would compromise the ability of the liver to mobilize glycogen stores.

486. The answer is C. *(Stryer, 4/e, pp 638–639.)* Pantothenic acid, which is a component of coenzyme A (CoA), must be obtained in the diet. It is synthesized by plants and microorganisms. Although niacin is a vitamin, it can be synthesized in humans from the essential amino acid tryptophan. During amino acid degradation, metabolic intermediates are produced and converted into glucose, fatty acids, or ATP in the citric acid cycle. In addition, all amino acids produce a urea nitrogen for elimination from the kidneys. Amino acids that form acetyl CoA are ketogenic and may give rise to ketone bodies or any derivative of acetyl CoA, such as fatty acids. Glucogenic amino acids form intermediates that may undergo gluconeogenesis to yield glucose. Sphingosine, the backbone of ceramides, is formed from serine and palmitoyl CoA.

487. The answer is C. *(Stryer, 4/e, pp 770–773.)* Although the liver is the major site of the formation of free glucose to maintain blood glucose levels, the kidneys and intestinal epithelium (e.g., duodenum, jejunum, and ileum) may also release glucose. All of these tissues contain the enzyme glucose-6-phosphatase. Glucose-6-phosphatase is an endoplasmic reticulum enzyme that dephosphorylates glucose and allows it to be transferred out of the cells. No other tissues in mammals contain this enzyme.

488. The answer is E. *(Stryer, 4/e, pp 314–316, 392–393.)* All the terms in the question refer to a striated muscle cell. The sarcolemma is the plasma membrane upon which nerve-muscle end plates impinge. The sarcoplasmic reticulum is the endoplasmic reticulum of the muscle cell. In addition to the normal reticulum activities, the endoplasmic reticulum is specialized to sequester calcium ions during relaxation and release them during contraction. Cardiac muscle cells are joined in an electrical syncytium by membranous attachments to each other called *intercalated disks.* The sarcomere is the functional unit of a muscle cell and is composed of a contractile unit of thick and thin filaments. The sarcoplasm is the cytoplasm of a muscle cell.

489–493. The answers are 489-C, 490-L, 491-G, 492-D, 493-K. *(Stryer, 4/e, pp 174–175, 196–199, 607–608, 813, 906, 941–942.)* A deficiency in carnitine, carnitine acyltransferase I, carnitine acyltransferase II, or acylcarnitine translocase can lead to an inability to oxidize long-chain fatty acids. This occurs because all of these components are needed to translocate activated long-chain (>10 carbons long) fatty acyl CoA across mitochondrial inner membrane into the matrix where β-oxidation takes place. Once long-chain fatty acids are coupled to the sulfur atom of CoA on the outer mitochondrial membrane, they can be transferred to carnitine by the enzyme carnitine acyltransferase I, which is located on the cytosolic side of the inner mitochondrial membrane. Acyl carnitine is transferred across the inner membrane to the matrix surface by translocase. At this point the acyl group is reattached to a CoA sulfhydryl by the carnitine acyltransferase II located on the matrix face of the inner mitochondrial membrane.

One of the primary killers of children prior to immunization was upper respiratory tract infections by *Corynebacterium diphtheriae.* Toxin produced by a lysogenic phage that is carried by some strains of this bacteria causes the lethal effects. It is lethal in small amounts because it blocks protein synthesis.

The viral toxin is composed of two parts. The B portion binds a cell's surface and injects the A portion into the cytosol of cells. The A portion ADP-ribosylates a histidine-derived residue of the elongation factor 2 (EF-2) known as *diphthamide*. This action completely blocks the ability of EF-2 to translocate the growing polypeptide chain.

Wood alcohol (methanol) is a cause of death or serious illness (including blindness) among patients who ignorantly substitute it for ethanol or mistakenly ingest it. Ingestion of automotive antifreeze (ethylene glycol) can also result in death if not treated. In both cases, death or serious injury can be averted by quickly administering an intoxicating dose of ethanol. The success of this treatment is based upon the fact that methanol and ethylene glycol are not poisons as such. First, they must be converted by the action of the enzyme alcohol dehydrogenase to precursors of potentially toxic substances. Administration of large doses of ethanol inhibits oxidation of both methanol and ethylene glycol by effectively competing as a preferred substrate for the active sites of alcohol dehydrogenase. Over time, methanol and ethylene glycol are excreted.

In many pathologies involving hemoglobin, the amounts of hemoglobin synthesized are normal, but problems are caused by mutations of a single amino acid residue. In contrast, thalassemias are a distinct class of genetic hemoglobin disorders distinguished by abnormal synthesis of the hemoglobin chains. In thalassemias, deficiencies can be due to abnormal RNA synthesis or processing, impaired protein synthesis, or loss of the globin gene itself.

The skin of patients with the autosomal recessive disease xeroderma pigmentosum is extraordinarily sensitive to light. Many patients die from metastases of malignant skin tumors. The genetic deficiency is a lack of a repair excinuclease responsible for hydrolyzing the DNA backbone near a pyrimidine dimer produced by ultraviolet radiation. Under normal circumstances, three enzymes participate in excision and repair of pyrimidine dimers. Initially a uvrABC-enzyme complex detects the pyrimidine dimer and its associated excinuclease cuts the DNA strand at specific points on each side of the dimer. (This enzyme is lost in xeroderma patients.) DNA polymerase I synthesizes a replacement strand and DNA ligase joins the 3′ end of the new stretch of DNA chain to the original strand.

494–496. The answers are 494-G, 495-A, 496-F. *(Stryer, 4/e, pp 765–777.)* During a time of need such as overnight fasts or starvation, hormone-sensitive lipase is activated by glucagon or epinephrine in adipose tissue. This leads to hydrolysis of triacylglyceride stores with the release of free fatty acids and glycerol into the blood. The free fatty acids are utilized by muscle as a preferred energy source and the glycerol is transformed into glycerol phosphate in the liver for reutilization in gluconeogenesis or glycolysis. Glucagon, epinephrine, and norepinephrine activate a cyclic AMP–dependent cascade in adipose tissue that ultimately leads to phosphorylation and activation of hormone-sensitive triacylglyceride lipase in adipose tissue. In contrast, dephosphorylation of the hormone-sensitive lipase leads to its inactivation. This occurs when adenylate cyclase is inhibited by the action of insulin. In contrast to hormone-sensitive triacylglyceride lipase, which regulates lipolysis in adipocytes, lipoprotein lipase is found on the surface of the plasma membranes of endothelial cells lining capillaries of many tissues. Lipoprotein lipase is activated by apolipoproteins found in the lipoproteins circulating in blood following the absorptive phase. Once activated, lipoprotein lipase leads to the breakdown of triacylglycerides found in very low-density lipoproteins and chylomicrons. The hydrolysis of these triacylglycerides leads to the formation of free fatty acids, which can enter adipose tissue and muscle tissue where they are either stored or used for energy. In liver, an increase in activity in the enzyme phosphofructokinase leads to a stimulation of glycolysis. This occurs in the well-fed state by an increase in fructose-2,6-bisphosphate, which is an allosteric activator of phosphofructokinase. In contrast, gluconeogenesis is inhibited by fructose-2,6-bisphosphate because it inhibits fructose-1,6-bisphosphate. Thus, under conditions that promote glycolysis, gluconeogenesis is inhibited, thereby preventing futile recycling of substrates and products.

497–500. The answers are 497-D, 498-B, 499-E, 500-C. *(Stryer, 4/e, pp 770–773, 775–780.)* In a normal, postabsorptive patient, blood fuel values are 4.5 mM glucose, 0.5 mM free fatty acids, 0.02 mM ketone bodies, and 4.5 mM amino acids. The ketone bodies are always low in a fed person.

Following several days of starvation, a catabolic homeostasis has set in, such that free fatty acids (1.5 mM) have risen and production of ketone bodies (5 mM) by the liver is proceeding. At this point, glycogen stores have been depleted, and much of the blood glucose, which will be maintained at about 4.5 mM throughout starvation, now comes from gluconeogenesis of amino acids (4.7 mM) derived from protein breakdown. Most of the brain's fuel supply still derives from glucose at this time. Since the brain accounts for at least 20 percent of the body's total consumption of fuel, this amount can be considerable.

Following prolonged starvation, utilization of glucose and hence catabolism of protein are spared by the induction of increased amounts of brain enzymes to utilize ketone bodies. Thus, in prolonged starvation, the blood concentration of amino acids (3.1 mM) decreases, whereas that of free fatty acids (2 mM) and ketone bodies (8 mM) increases. Of course, blood glucose is maintained at about 4.5 mM.

The lack of insulin in diabetics causes a stimulation of lipolysis, glycogenolysis, gluconeogenesis, and ketogenesis. Thus, the blood values of free fatty acids (2 mM), ketone bodies (10 mM), and amino acids (4.5 mM) should resemble those of a fasting or starving person with one major exception—the high level of blood glucose (12 mM). The lack of insulin does not allow the glucose to enter most cells.

Bibliography

Rawn JD: *Biochemistry.* Burlington, NC, Neil Patterson, 1989.

Stryer L: *Biochemistry,* 4/e. New York, Freeman, 1995.